# Numerical and Engineerin

## Computer-Aided Design by Python

# Numerical and Engineering Analysis

## Computer-Aided Design by Python

Dr. Samir Abood
Dr. Naima Khatir
Dr. Muna Fayyadh

*CWP*

**Central West Publishing**

**Disclaimer**
Every effort has been made by the publisher, editors and authors while preparing this book, however, no warranties are made regarding the accuracy and completeness of the content. The publisher, editors and authors disclaim without any limitation all warranties as well as any implied warranties about sales, along with fitness of the content for a particular purpose. Citation of any website and other information sources does not mean any endorsement from the publisher, editors and authors. For ascertaining the suitability of the contents contained herein for a particular lab or commercial use, consultation with the subject expert is needed. In addition, while using the information and methods contained herein, the practitioners and researchers need to be mindful for their own safety, along with the safety of others, including the professional parties and premises for whom they have professional responsibility. To the fullest extent of law, the publisher, editors and authors are not liable in all circumstances (special, incidental, and consequential) for any injury and/or damage to persons and property, along with any potential loss of profit and other commercial damages due to the use of any methods, products, guidelines, procedures contained in the material herein.

NATIONAL LIBRARY OF AUSTRALIA

A catalogue record for this book is available from the National Library of Australia

ISBN (print): 978-1-922617-40-8

# Preface

The book is designed for use in a graduate program in Numerical Analysis that includes a basic introductory course and subsequent more specialized courses. The latter is envisaged to cover numerical linear algebra, the numerical solution of ordinary and partial differential equations, and perhaps additional topics related to complex analysis, multidimensional analysis, in particular optimization, and functional analysis and related functional equations.

Viewed in this context, the first four chapters of our book could serve as a text for the basic introductory course on the Python program, and the remaining chapters could provide a text for an advanced course on the numerical solution of ordinary differential equations. In a sense, therefore, the book breaks with tradition in that it no longer attempts to deal with all major topics of numerical mathematics. Those dealing with linear algebra and partial differential equations have developed into major fields of study that have attained a degree of autonomy and identity that justifies their treatment in separate books and separate courses on the graduate level. The term "Numerical Analysis" as used in this book, therefore, is to be taken in the narrow sense of the numerical analog of Mathematical Analysis, comprising such topics as machine arithmetic, the approximation of functions, approximate differentiation and integration, and the approximate solution of nonlinear equations and ordinary differential equations.

This book aims to provide a good understanding of Numerical engineering analysis and its applications and optimization. The book begins with studying the concept of Python fundamentals for scientific computing. It then presents their applications in the different configurations shown in lucid detail.

It optimizes the protective scheme's location in power and uses the power electronics devices and install to protect the power system. This book is intended for college students, community colleges and universities. The book is also intended for researchers, technicians, and technology and skills specialists in the power and control of power systems. This book presents the relation between the power system's quantities and their protection and management. The book's major goal is to give a concise introduction engineering analysis covered in two semesters. The book is appropriate for Juniors, Senior Undergraduate Students, Graduate Students, Researchers, and Academics.

This book is organized into 10 chapters. Chapter 1 introduces the Python fundamentals for scientific computing, the Execution of a first program, and Some quick basics in Python. In recognition of requirements by the Accreditation Board for Engineering and Technology (ABET) on integrating computer tools, Python is encouraged in a student-friendly manner. The reader does not need to have previous knowledge of Python. The material of this text can be learned without Python. However, the authors highly recommend that the reader studies this material in conjunction with the Python Student Version.

Chapter 1 of this text provides a practical introduction to Python.

Chapter 2 concerns some aspects of operations with the complex number and Polar form of complex numbers; also, the chapter discusses the Euler's Formula and De Moivre's formula. This chapter discusses the complex function, Complex Integral, and Basic properties of line integrals. The chapter is focused on Cauchy's Integral theorem and Cauchy 's Integral Formula.

Chapter 3 describes the preceding of the Simple Matrix Algebra, Matrix Multiplication, and properties of matrix multiplication. Block Multiplication, Matrix Inverses, and Matrix Inversion by Gaussian Elimination in this chapter.

Chapter 4 deals with Matrix operations and the order of a matrix. The system of linear equations, Eigen values, Eigen vectors, and LU-Decomposition for Band Matrices is also discussed.

Chapter 5 presents the Optimization and form Changing of Optimization Problem. This chapter also studies Conventional Form, Different Kinds of Optimization, Golden-Section Search, and Newton's method. The description of Numerical Integration and Numerical solution of differential equations are discussed in chapter 6 under the title of Numerical integration. The chapter includes an introduction to the rectangular rule and the Trapezoidal rule. Also, the chapter discusses Euler's method, Richardson h2 extrapolation and Rung–Kutta method.

Chapter 7 begins with the definition of the Laplace transform and uses that definition to derive the transform of some basic, important functions. This chapter considers some properties of Laplace transform, which help obtain the Laplace transform of other functions, then consider the inverse Laplace transform.

Chapter 8 discusses the z-transform, region of convergence (roc), properties of the z-transform, also discusses the inverse of z-transform. Chapter 9 describes the z-transform applications as an evaluation of LTI system response using z-transform and implementation using z-transform.

Finally, Chapter 10 introduces pole-zero stability, Difference Equation and Transfer Function, and the Stability of the systems.

# About the Authors

*Samir I. Abood* received his BS and MS from the University of Technology, Baghdad, Iraq, in 1996 and 2001; respectively, he got his Ph.D. in the Electrical and Computer Engineering Department / Prairie View A & M University. From 1997 to 2001, he worked as an engineer at the same university. From 2001 to 2003, he was a professor at the University of Baghdad and Al-Nahrain University. From 2003 to 2016, Mr. Abood was a Middle Technical University / Baghdad – Iraq professor. From 2018 to the present, he has worked at Prairie View A & M University/ Electrical and Computer Engineering Department. He is the author of 30 papers and seven books. His main research interests are sustainable power and energy systems, microgrids, power electronics and motor drives, digital PID controllers, digital methods for electrical measurements, digital signal processing, and control systems.

*NAIMA Khatir*, born in 1982, is a researcher at LTE laboratory and lecturer at the university center of Naama (Algeria). He obtained his Ph.D. in mechanical engineering in 2015. His research works focus on the combustion performances and emissions of an internal combustion engine. He worked on Hydrogen fuelled spark ignition engine and has many publications relating to this topic. Currently, NAIMA Khatir continues his research in diesel engines running with alternative fuels. He has many papers published in this area, and other papers are in progress.

*Muna H. Fayyadh* has received her BS and MS from Al- Nahrain University, Baghdad, Iraq, in 1998 and 2002, respectively. She got her Doctoral degree in computer science - Cybersecurity concentration in 2019 from Colorado Tech University. She has worked as an instructor at Al-Nahrain university from 2001-2005. She worked as an IT specialist at the Ministry of water resources in Iraq from 2003-2013. From 2014-2018 she worked as an IT instructor for a private school in Houston, Tx. From 2018 -2020 she worked at the City of Houston, an adjunct professor (Grad level) at American Intercontinental University and Houston community college. She is the author of 15 papers and three books. Her main research interests are cybersecurity and intrusion detection techniques, network security, machine learning, and AI.

# Contents

# Chapter 1

## Python Fundamentals for Scientific Computing

In this introduction to Python for scientific programming, we will be using Python as a language and the NumPy and Matplotlib libraries for scientific computation and visualization. To run the programs in this course on your computer, you will need Python and the NumPy and Matplotlib libraries.

You can install the free Anaconda software on your computer, which provides easy installation of the entire development environment required for this course:

**https://www.anaconda.com/download/**

### 1.1 Keywords

Keywords are predefined and reserved words used in programming which have special meaning for the compiler. Keywords are part of the syntax and cannot be used as identifiers.

There are 33 keywords in Python 3. This number may vary slightly over time.

All keywords except True, False, and None are lowercase and should be written as is. The list of all keywords is given below.

| FALSE | class | finally | is | return |
|--------|----------|---------|----------|--------|
| None | con-tinue | for | lambda | try |
| TRUE | def | from | nonlocal | while |
| and | del | global | not | with |
| as | elif | if | or | yield |
| assert | else | import | pass | |
| break | except | in | raise | |
| | | | | |

The identifier is the name given to entities such as variables, functions, classes, etc.

Identifiers must be unique. They are created to give a unique name for an entity to identify it during program execution.

## 1.2 Execution of a first program

Here is the first example of a program:

```
for i in range (5):
    print("Value of i is:", i)
print("end of the loop")
```

This program is a loop that displays the variable value of i from 0 to 5, and it should be noted that the syntax "in range i" means the incrementation of the value from 0 to 4. The display after execution is:

```
Value of i is: 0
Value of i is: 1
Value of i is: 2
Value of i is: 3
Value of i is: 4
end of the loop
```

## 1.3 Some quick basics in Python

Use in interactive mode

In Spyder, we have a window area for the IPython command interpreter. Its command prompt is different from that of the standard Python interpreter (which is>>>). It looks like this.

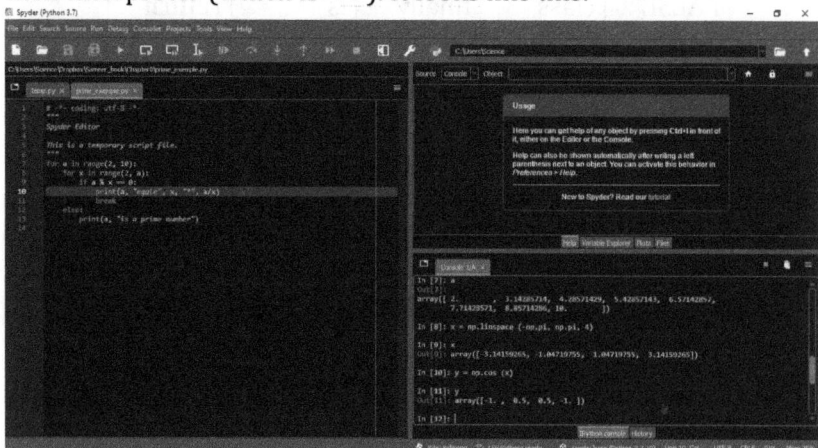

Example of simple calculation

```
In [1]: 4 * 5
Out[1]: 20

In [2]: 6 + 4
Out[2]: 10
```

The parts that start with In and Out correspond to the inputs and outputs respectively.

First calculations

Test the following calculations:

```
4 + 5
>>> 3 - 7
>>> 5 + 2 * 3
>>> (6 + 3) * 2
```

**To calculate a power, we use \*\*.**

```
4**3
```

**The operator/**
In Python 3, The operator / performs a standard decimal division like on a calculator

```
5 / 2
2.5
```

**To perform an integer division, use //:**

```
5 // 2
```

## The operator %

The% operator (called the modulo operator) provides the remainder of the integer division of one number by another.

Example

```
5 % 2
1
>>> 8 % 4
0
```

## Assignment

```
>>>
a = 3
>>> a
3
>>> b = a + 4
>>> b
7
```

The first line contains the instruction a = 2. To understand this instruction, one must imagine that the information is stored in boxes within the computer's memory. To manipulate the information, we give names to these boxes. Here we create a box called a, and we assign the value 2. We will talk about variable a since a can contain variable values in the following. In other words, the instruction a = 2 is an assignment instruction that puts the value 2 in the variable a.

Display – the function print ()

To display, we use the **print () function**.

```
>>> print("Numerical methods with Python")
```

Numerical methods with Python

```
>>> a = 32
>>> print(a)
 32
```

It is possible to carry out several displays in succession. To do this, we separate the elements with commas.

```
>>> a = 5
>>> print("The value of a is:", a)
The value of a is: 5
```

The function range ()
If you need to create a series of integers, you can use the **range ()**
**function**. It generates an arithmetic sequence.

```
>>> range(10)
range (0, 10)
```

In Python 3, if you want to display the values, you need to convert the result into a list using the list () function. For example:

```
>>> list(range(10))
[0, 1, 2, 3, 4, 5, 6, 7, 8, 9]
```

The ending number passed to it is never in the generated list. For example, range (10) generates ten values, exactly the indices of the elements of a sequence of length 10. It is possible to start the interval at another number or to specify a different increment (even negative):

```
> >> list(range(15, 20))
[15, 16, 17, 18, 19]
>>> list(range(0, 10, 3))
[15, 17, 19]
>>> list (range ( -20, -100, -10))
[-20, -30, -40, -50, -60, -70, -80, -90]
```

In general, we have:

range (initial_value, end_point, step)

The step can be positive or negative. The value of the end limit (maximum or minimum) is never reached.

Exercise

Create a list containing the even integers going from 10 to 40.

## 1.4 Access to the elements of a list

```
>>> a = list (range (2,20,2))
>>> a
 [2, 4, 6, 8, 10, 12, 14, 16, 18]
>>> a[0]
 2
>>> a[2]
 6
```

To access an element of a list, one indicates between brackets [] the index of the element.

The function len ()

lenfunction()returns the number of items. For example:

```
>>> a = a = list (range (5,10))
>>> a
[5, 6, 7, 8, 9]
>>> len(a)
5
```

**Reading information from the keyboard**

the function input ()

```
var = int(input("give an integer:"))
```

By default, the function **input** ()returns a string. Therefore, we must use the function int (), which allows us to obtain an integer.

**Another example of program**

```
n = 5
print("We will ask to give an integer", n, "numbers")
for i in range(n):
    x = int(input("Give a number:"))
    yew x > 0:
        print(x, "is positive")
    else:
        print(x, "is negative or zero")
print("end")
```

To repeat: the **for**

```
for statement i in range(n):
    block Instruction
```

To test: thestatement**if**

```
if x > 0:
    print(x, "is positive")
else:
    print(x, "is negative or zero")
```

The different kinds of instruction
- simple
- instructions compound instructions like **if** or **for**
- instruction blocks

General writing rules

### *Identifiers*
An identifier is a series of characters used to designate the different entities handled by a program: variables, functions, classes...
In Python, an identifier is made up of letters or numbers. It does not contain space. The first character must be a letter. It can contain the character "_" (underscore, in French "underlined"). It is case-sensitive (the distinction between upper case and lower case).

### Keywords

Words reserved by the Python language (**if**, **for**, etc.) cannot be used as identifiers.

### Comments

The usual comments:

```
# This is a comment
```

Comments end of the line:

```
a = 2 # This is also a comment
```

## 1.5 Loops

Loops are used to repeatedly repeat the execution of part of the program.

Bounded and unbounded loops

**BoundedBounded loop**

> When we know how many times the repetition should occur, we usually use aloop**for**.

**Unbounded loop**

> If we do not know the number of repetitions in advance, we choose aloop**while**.

**Loop**

Example of use:

```
for i in [0, 2, 4, 6]:
    printPrint("the value of i is:", i)
```

after execution:

```
the value of i is:  0
the value of i is:  1
the value of i is:  2
the value of i is:  3
```

The statement **for** is a compound statement, that is, a statement whose header ends with a colon:, followed by an indented block that forms the body of the loop.

We say that we iterate the loop each time the body of the loop is executed.

In the header of the loop, we specify after the keyword **for** the name of a variable (i in the example above) which will successively take all the values which are given after the keyword **in**. We often say that this variable (here i) is a counter because it is used to number the iterations of the loop.

It is possible to obtain the same result without giving the list of values but using the **range () function**.

```
for i in range(4):
    print("the value of i is:", i)
```

the execution of this code will display the values of i from 0 to 3.
To browse the indices of a list, it is possible to combine **range ()** and **len ()** as below:

```
L = ["Book", "Learn", "Numerical", "Methods", "Python"]
for i in range(len(L)):
    print("i is", i, "and L [", i, "] is", L[i])
```

Display after execution:

```
i is 0 and L [0] is Book
i is 1 and L [1] is Learn
i is 2 and L [2] is Numerical
i is 3 and L [3] is Methods
i is 4 and L [4] is Python
```

Reminder:Thefunction **len ()** returns the number of elements:

```
>>> L = ["Book", "Learn", "Numerical", "Methods", "Python"]
>>> len(L)
```

In the following example, we will illustrate that the variable indicated after **for** goes through all the values of the list given after **in**:

```
L = ["Book", "Learn", "Numerical", "Methods", "Python"]
for i in c:
    print("i is", i)
>>>
i is Book
i is Learn
i is Numerical
i is Methods
i is Python
```

Loop **while**
Syntax:

```
while condition:
    Instruction A
```

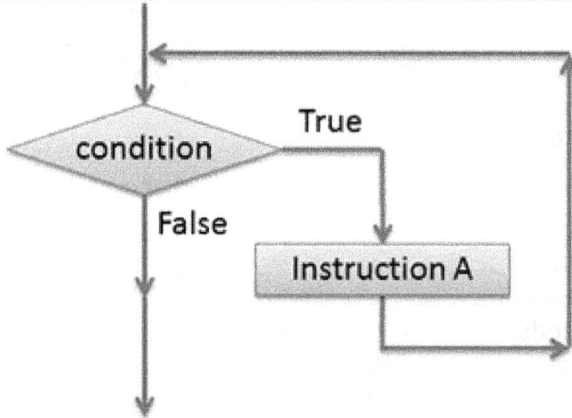

Example program:

```
a = 0
while at < 8:
```

```
  print("a is", a)
  at = at + 2
print("end of this loop")
```

Display after execution:

```
a is 0
a is 2
a is 4
a is 6
end of this loop, the body of the loop (i.e., the block of indented in-
structions ) will be repeated as long as the condition is true.
```

In the example above,x will be multiplied by 2 as long as its value is less than 10.
Note: If the condition is initially false, the body of the loop is never executed. If the condition still remains true, then the body of the loop is repeated indefinitely.
How to choose between loop **for** and loop **while**
In general, if we know before starting the loop the number of iterations to execute, we choose a loop **for**. On the contrary, if the decision to stop the loop can only be made by a test, we choose aloop **while**.

**Note**
It is always possible to replace aloop **for** with aloop **while**.
**Transforming aloop for into aloop** while

```
for i in range(5):
   print("the value of i is:", i)
```

The above program is equivalent to:

```
i = 0
while i < 5:
   print("the value of i is:", i)
   i = i + 1
```

**break** and **continue statements**, and theclause **else** in loops
break
The statement is **break** used to "break" the execution of a loop (**while** or **for**). It exits the loop and goes to the next statement.
Example

```
for i in range(10):
    print("start iteration", i)
    print("hello")
    yew i == 2:
        break
    print("end of iteration", i)
print("after loop")
```

Display after execution:

```
start of iteration 0
Hello iteration 0
end of iteration 0
start of iteration 1
Hello iteration 1
end of iteration 1
start of iteration 2
Hello iteration 2
End of the loop
```

**Note**
In the case of nested loops, the statement **break** exits only the inner-most loop.
**Note equivalent of do ... while (do ... while)**
In many languages, there is a do ... while statement, which allows you to create a loop for which you do not know the number of repetitions in advance, but which must be s 'run at least once. This instruction does not exist in Python, but we can easily reproduce its operation as follows:

```
while True:
```

```
   not = int(input ("Give a positive number:")
   print("you provided", n)
   yew not > 0:
      break
print("correct answer")
```

continue function
The instruction **continue allows** you to jump prematurely to the next loop turn. It continues on the next iteration of the loop.
Example

```
for i in range (4):
   print ("start of iteration", i)
   print ("iteration", i, "confirmed")
   if i <2:
      continue
   print ("Continue function is activated at iteration", i)
print ("End of Example")
```

Display after execution:

```
start of iteration 0
iteration 0 confirmed
start of iteration 1
iteration 1 confirmed
start of iteration 2
iteration 2 confirmed
Continue function is activated at iteration 2
start of iteration 3
iteration 3 confirmed
Continue function is activated at iteration 3
End of example
```

**The clause** else **in a loop**


Theclause**else**in a loop allows you to define a block of instructions which will be executed at the end only if the loop has been completed without being interrupted by a**break**.

13

Unlike the statements present after the loop, which are executed in all cases (with or without interruption by **break**), the block of statements defined in the clause will **else** not be executed during the interruption by **break**. After the interruption, we will go directly to the instructions after the loop.

In other words, the block of theclause **else** is executed when the loop ends with exhaustion of the list (with **for**) or when the condition becomes false (with **while**), but not when the loop is interrupted by **break**. This is illustrated in the following loop, which searches for prime numbers:

```
for at in range(2, 10):
    for x in range(2, a):
        yew at % x == 0:
            print(a, "equal", x, "*", a/x)
            break
    else:
        print(a, "is a prime number")
```

Display after execution:

```
2 is a prime number
3 is a prime number
4 equal 2 * 2.0
5 is a prime number
6 equal 2 * 3.0
7 is a prime number
8 equal 2 * 4.0
9 equals 3 * 3.0
```

**Introduction to Numpy**

The library **NumPy** (http://www.numpy.org/) allows performing numerical calculations with Python. It introduces easier management of the numbers, tables, vectors, matrices...etc

To use **NumPy**, it should be called at the environment that includes this library. You must first import the package **numpy** with the following instruction:

```
import numpy as np
```

Predefined
variables Pi variable
NumPy is used, for example, to obtain the value of pi.

```
>>> np.pi
3.141592653589793
```

tables -**numpy.array ()**
Creating
Tables can be created with **numpy.array**().Square brackets are used
to delimit lists of elements in arrays.

```
>>> x = np.array ([9, 8, 7, 6])
```

Display

```
>>> x
array ([9, 8, 7, 6])
>>> type(x)
numpy.ndarray
```

We see that we have obtained an object of type **numpy.ndarray**.
Accessing array elements

```
>>> x [0]
9
>>> x [3]
6
```

2D
array It is possible to create a 2D array using a list of lists with nested
brackets. Internal lists correspond to rows in the table.

15

```
>>> y = np.array ([[0, 2, 4], [6, 8, 10]])
```

Display

```
>>> y
array ([[0, 2, 4],
    [6, 8, 10]])
>>> type( y)
numpy.ndarray
>>> len(y) # Length of the array y
2
```

Access to the elements of a 2D array
We use a syntax with two indices. The first index is the index of the table row.

```
>>> b [0,1]
2
>>> b [1,2]
10
```

The function **numpy.arange ()**

```
>>> a = np.arange (2, 14, 2)
>>> a
array([ 2, 4, 6, 8, 10, 12])
>>> type(a)
numpy.ndarray
>>>len(a)
6
```

Note the difference between**numpy.arange ()**and**range ( )**:
* **numpy.arange ()** **a fucntion** returns an object of type **numpy.ndarray**. it generates numbers that are acceded as a numbers stored in a numpy array.

- **range ()** is a built-in function that returns an object of type **range**. It generates integers that can be treated as a list.

The function **numpy.linspace ()**

**numpy. linspace ()**allows obtaining a 1D array going from a starting value to an ending value with a given number of elements. Here in this example, we want to create an array from 2 to 10 with 10 elements:

```
>>> a= np.linspace (2, 10, 10)
array([ 2.    , 2.88888889, 3.77777778, 4.66666667, 5.55555556,
    6.44444444, 7.33333333, 8.22222222, 9.11111111, 10.    ])
```

The action of a mathematical function on an array

NumPy has a large number of mathematical functions that can be applied directly to an array. In this case, the function is applied to each of the elements of the array.

```
>>> x = np.linspace (-np.pi, np.pi, 4)
>>> x
array([-3.14159265, -1.04719755, 1.04719755, 3.14159265])
>>> y = np.cos (x)
>>> y
array([-1. , 0.5, 0.5, -1. ])
```

Get 2D array row or column¶

```
>>> a = np.arange (1, 6)
>>> a
array ([1, 2, 3, 4, 5])
>>> a.shape = (1, np.size (a))
>>> a
array ([[1, 2, 3, 4, 5]])
>>> a.shape = (np.size (a), 1)
>>> a
array ([[1],
    [2],
    [3],
```

```
    [4],
    [5]])
```

## Calculations on matrices

### Matrix generation

To create a matrix using Python, we should import NumPy using the flowing command:
import numpy as np
let consider the following matrix A :

1   2   3

4   5   6

7   8   9

The matrix has 3 rows and 3 columns. In Python, such matrix is crated as follows:
A = np.array([[1, 2, 3],[4, 5, 6],[7, 8, 9]])
In the same manner, we can create a vector C

$$C = \begin{pmatrix} 12 \\ 8 \\ 10 \end{pmatrix}$$

C = np.array([12,8,10])

## Operations on matrices

### Matrix Addition

To perform addition on the matrix, we will create another matrix B using numpy.array() and add them using the (+) operator.

−3   1   2

6   −4   5

8   9   −2

Matrix B is created as follows:
B = np.array([[-3, 1, 2],[6, -4, 5],[8, 9, -2]])
And the addition is performed using (+)
M1 = A + B
print(M1)

The output is:
[[-6 2 4]
 [12 -8 10]
 [16 18 -4]]
In the same manner, the subtraction can be performed
M2 = A - B
print(M1)

## Matrix Multiplication

To multiply two matrices with the numpy we can make use of dot()
instruction. The dot product of matrices M1 and M2 is Numpy. dot().
Numpy.dot() is responsible for 2D arrays and matrix multiplications.
M3 = np.dot(A,B)
print(" The multiplication of A and B is M2=", M2)
 The output is:
The multiplication of A and B is
M2= [[ 31 11 -5]
  [ -2 67 -18]
  [ 14 -46 65]]

## Transpose, inversion and determinant of matrix

# Transpose
M3 = np.transpose(A)
print(" the transpose of is M3=", TransA )
#Determinant DetA
M4 = np.linalg.det(A)
print("The determinent of matrix A is M4:", DetA)
#inverse
M5 = np.linalg.inv(A)
print("The inverse of matrix A is M5:", M5)
**The output is:**
The transpose of is
M3= [[-3 6 8]
  [ 1 -4 9]
  [ 2 5 -2]]
The determinent of matrix A is: 334.99999999999994
The inverse of matrix A is M5:
[[-0.11044776 0.05970149 0.03880597]
 [ 0.15522388 -0.02985075 0.08059701]

[ 0.25671642 0.10447761 0.01791045]]
**solution of linear equations using NumPy**
**we consider the following system of linear equations:**

$$\begin{pmatrix} 1 & 2 & 3 \\ 4 & 5 & 6 \\ 7 & 8 & 9 \end{pmatrix} \cdot \begin{bmatrix} x \\ y \\ z \end{bmatrix} = \begin{bmatrix} 12 \\ 8 \\ 10 \end{bmatrix}$$

**The script to solve this system is:**
**# Solution of system of linear equations A.X=C**
**M5 = np.linalg.solve(A,C)**
**print("The solution of A.X=C is M5: ", M5)**
**print(" X=", M5[0], " Y=", M5[1], " Z=", M5[2])**
**the output is :**
**The solution of A.X=C is M5: [-0.45970149 2.42985075 4.09552239]**
**X= -0.4597014925373132 Y= 2.4298507462686567 Z= 4.095522388059702**

### 1.6 Plotting curves

To draw curves, we need the libraries **NumPy** and **matplotlib** used in this course.
For the "standard" syntax, you must import the package numpy and the module pyplot from **matplotlib**. We must then specify the libraries when calling functions (see the examples below
Creating a curve
Using **plot ()**
The instruction **plot ()** is used to draw curves which connect points whose abscissas and ordinates are provided in tables

### Example 1.1

```
import numpy as np
import matplotlib.pyplot as plt

x = np.array ([1, 2, 3, 5])
y = np.array ([2, 4, 6, 10])
plt.plot (x, y)

plt.show () # displays the figure on the screen
```

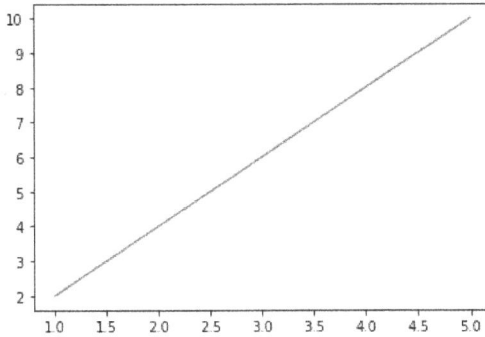

Example 1.2

In this example, we are going to draw the curve of sin (x). x varies from - pi a + pi with 50 points in interval. As the example it is necessary to call the library numpy and **numpy** and **matplotlib.** The function plt.plot (t, y) allows to create the tras while plt.show (t, y) used to display the curve on the screen.

```
import numpy as np
import matplotlib.pyplot as plt
x = np.linspace (-np.pi, np.pi,50)
y = np.sin (x)
print(y)

plt.plot (t, y)
plt.show (t, y)
```

program execution will display the following figure:

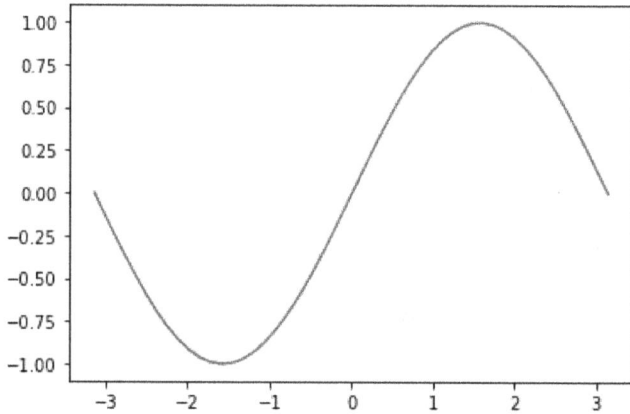

Adding a title -title ()
You can add a title using the instruction **title ()**.

```
import numpy as np
import matplotlib.pyplot as plt

x = np.linspace (-np.pi, np.pi, 50)
y = np.sin (x)

plt.plot (x, y)

plt.title ("Sine function")

plt.show ()
```

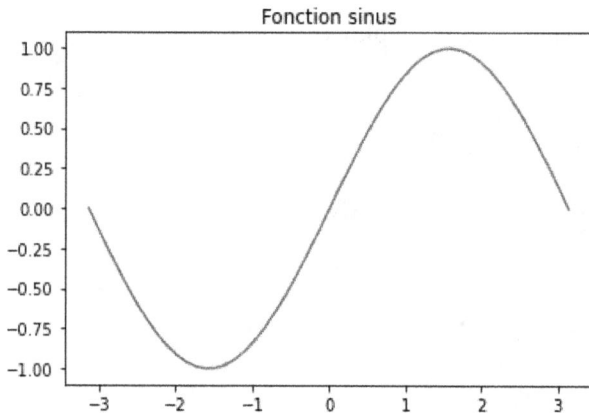

Fonction sinus

Adding a legend -legend ()

```
import numpy as np
import matplotlib.pyplot as plt
x = np.linspace (-np.pi, np.pi,50)
y = np.sin (x)
print(y)

plt.plot (x, y, label="cos (x)")
plt.legend ()
plt.title ("Sine function")
plt.show (x, y)
```

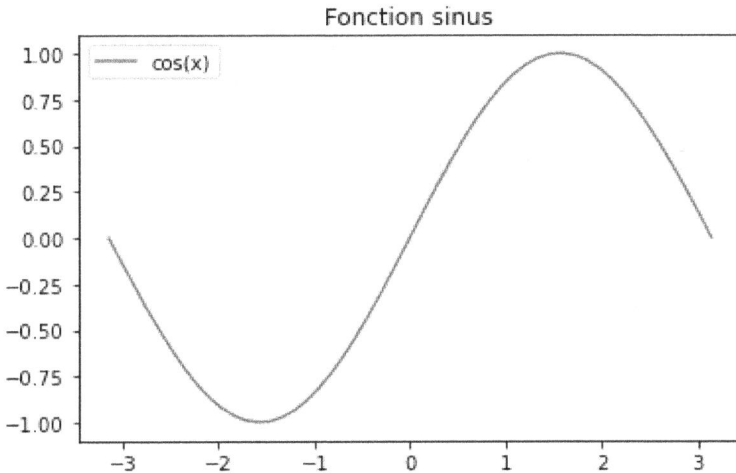

Labels on the axes -xlabel ()andylabel ()
Labels on the axes can be added with the functions**xlabel ()**and**yla-bel ()**.

```
import numpy as np
import matplotlib.pyplot as plt
x = np.linspace (-np.pi, np.pi,50)
y = np.sin (x)
print(y)

plt.plot (x, y, label="cos (x)")
plt.legend ()
plt.title ("Sine function")
plt.xlabel ("Values of X")
```

plt.ylabel ("Values of sin (x)")
plt.show (x, y)

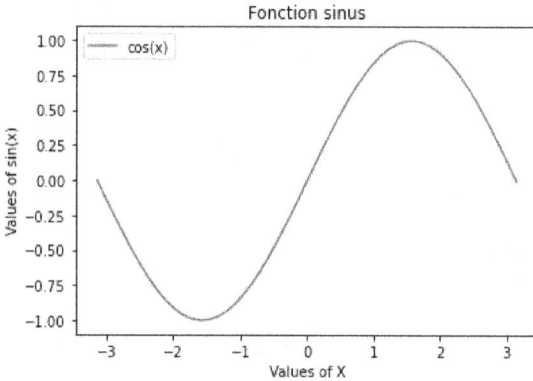

Fonction sinus

## Exercise 1
Predict what will be displayed when the following python commands are executed. Write them and explain who they work. Use help resources as needed.

**import numpy as np**

A = np.matrix (' 3,6,9 ; 2,5,8; 1,4,7')
print (A)

B = np.array([[1,2,3], [4,5,6], [7,8,9]])
print(B)

G = np.matrix (' 3 4 -1 ; -2 5 2; -1 4 5')
print (G)

Write the python script that calculate the following operation:
- A+B, A-B, A*B
- Transpose of A and B
- Determinant of A and B

## Exercise 2
Create a script program that create the two vectors a and b, than calculate the product operations a*b and a (dot) b, use the command np.dot(a,b). compare between the two results.

24

a = ( 1 , 3, 5 )
b = ( 3, 1, 2)

**Exercise 3**
Create using commands np.zeros and np.ones a matrix *4x3* for each function.

**Exercise 4**
Use linspace function to create a vectors equivalent to the following python statements:
C = np.arange(0,35,5)
E = np.arange(-5,10)

**Exercise 5**
Instead of linspace use numpy arrange function to create vectors equivalent to the following Python statements:
F = np.linspace(-4, 8, 10)
print(F)
H = np.linspace(-4, 4, 5)
print(H)

**Exercise 6**
Generate a plot of y= cos(x), include a title, axis labels, and a legend for x varies from $[\pi/4, 3\pi/4]$.

**Exercise 7**
Repeat exercise 6 with the equation $y = cos(x^2+3x) + 4x^2-3$ where x varies from [0 to 50] with 100 points.

**Exercise 8**
The temperature dependency of the rate of reaction k and the activation energy E is given by the following relation:
$$k = Ae^{-Ea/(RT)}$$
where A is the pre-exponential factor for the reaction, R is the universal gas constant *(8.314 J/mol.K)*, T is the absolute temperature (usually in kelvins), and k is the reaction rate coefficient.
For *k= 7.1016 (1/s), E=105 J/mol*
Use Python to create calculate the reaction rate with respect to temperature in a range of temperature from 250 to 425 (without using a loop).
The curve should look like this

25

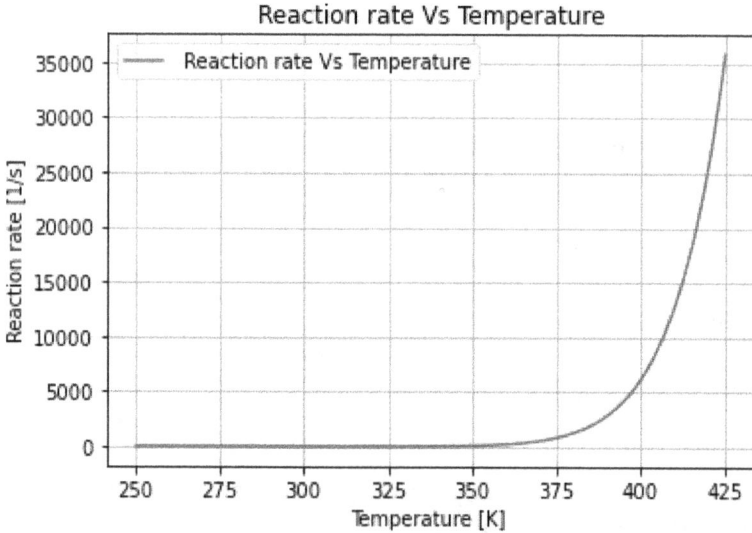

Reaction rate Vs Temperature

**Solution**

```
import numpy as np
import matplotlib.pyplot as plt

k0= 7e16
E = 1e5
R = 8.314
T = np.linspace(250,425,100)

k = k0 * np.exp(-E/R/T)


plt.plot(T,k, label=" Reaction rate Vs Temperature")
plt.legend()
plt.title("Reaction rate Vs Temperature")
plt.xlabel("Temperature [K] ")
plt.ylabel("Reaction rate [1/s]")
plt.show
plt.grid()
```

# Chapter 2

# Complex numbers

The complex number is usually denoted by zwhere"z" is defined in terms of an ordered pair of real number $(x, y)$ and imaginary number $j = \sqrt{-1}$ such that Z

$$Z = x + jy$$

Where

    x is the real part of Z:R(Z)

    y is the imaginary part of Z:I(Z)

Note that :

$$j = \sqrt{-1}, j^2 = -1, j^3 = -j$$
$$j^{-1} = \frac{1}{j} = -j, j^{-2} = \frac{1}{j^2} = -1, j^{-3} = \frac{1}{j^3} = \frac{1}{-j} = +j. \text{ Since } j = \sqrt{-1}$$
$$\frac{1}{j} = -j$$
$$j^2 = -1$$
$$j^3 = j \cdot j^2 = -j$$
$$j^4 = j^2 \cdot j^2 = 1$$
$$j^5 = j \cdot j^4 = j$$

A second way of representing the complex number z is by specifying it magnitude r and angle $\theta$ it makes with the real axis, as shown in Figure 2.1. This is known as the polar form. It is given by

$$z = |z|\angle\theta = r\angle\theta \qquad (2.4)$$

where

$$r = \sqrt{x^2 + y^2}, \qquad \theta = \tan^{-1}\frac{y}{x} \qquad (2.5a)$$

or

$$x = r\cos\theta, \qquad y = r\sin\theta \qquad (2.5b)$$

that is,

$$z = x + jy = r\angle\theta = r\cos\theta + jr\sin\theta \qquad (2.6)$$

In converting from rectangular to polar form using Equation (2.5), we must exercise care in determining the correct value of $\theta$. These are the four possibilities:

$$z = x + jy, \qquad \theta = \tan^{-1}\frac{y}{x} \qquad \text{(1st quadrant)}$$

$$z = -x + jy, \qquad \theta = 180° - \tan^{-1}\frac{y}{x} \qquad \text{(2nd quadrant)}$$

$$z = -x - jy, \qquad \theta = 180° + \tan^{-1}\frac{y}{x} \qquad \qquad \text{(3rd quadrant)}$$

$$z = x - jy, \qquad \theta = 360° - \tan^{-1}\frac{y}{x} \qquad \qquad \text{(4th quadrant)}$$

$$(2.7)$$

**Imaginary**

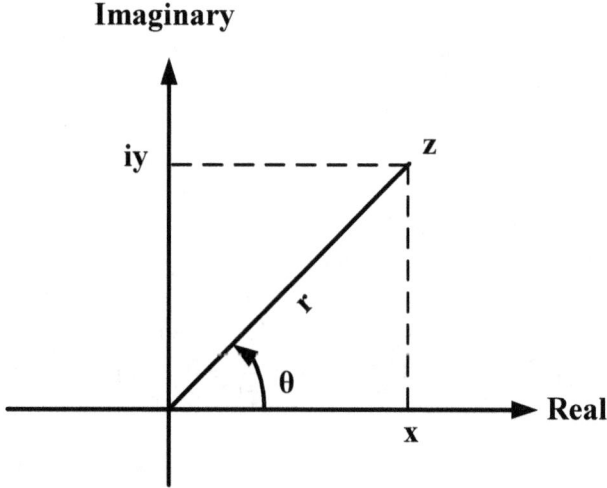

Figure 2.1 Graphical representation of a complex number.

if x and y are positive.

The third way of representing the complex number x is the exponential form:

$$z = re^{j\theta} \qquad\qquad (2.8)$$

This is almost the same as the polar form, because we use the same magnitude r and the angle $\theta$.

The three forms of representing a complex number are summarized as follows.

$$z = x + jy, \qquad (x = r\cos\theta, y = r\sin\theta) \qquad \text{Rectangular form}$$

$$z = r\angle\theta, \qquad (r = \sqrt{x^2 + y^2}, \theta = \tan^{-1}\frac{y}{x}) \quad \text{Polar form}$$

$$z = re^{j\theta}, \qquad (r = \sqrt{x^2 + y^2}, \theta = \tan^{-1}\frac{y}{x}) \quad \text{Exponential form}$$

$$(2.9)$$

## 2.1 Operations with complex number

if $Z_1 = x_1 + jy_1$ and $Z_2 = x_2 + jy_2$ are two complex number then:

a-     $Z_1 = Z_2$ if $x_1 = x_2$ and $y_1 = y_2$

b-

$$Z_1 + Z_2 = (x_1 + jy_1) + (x_2 + jy_2) = (x_1 + x_2) + j(y_1 + y_2)$$
$$Z_1 - Z_2 = (x_1 - x_2) + j(y_1 - y_2)$$

c-     $Z_1 . Z_2 = (x_1 + jy_1)(x_2 + jy_2) = (x_1 x_2 - y_1 y_2) + j(x_1 y_2 + x_2 y_1)$

d-     Conjugate    of    $Z_1 = x_1 + jy_1$    is    $\overline{Z_1} = x_1 - jy_1$

$$Z_1 \overline{Z_1} = x_1^2 + y_1^2 \quad , Z_1 + \overline{Z_1} = 2x_1 \quad , Z_1 - \overline{Z_1} = 2y_1 j$$

$$\overline{Z_1 \pm Z_2} = \overline{Z_1} \pm \overline{Z_2}$$

$$\overline{Z_1 . Z_2} = \overline{Z_1} . \overline{Z_2}$$

e-

$$\frac{Z_1}{Z_2} = \frac{x_1 + jy_1}{x_2 + jy_2} = \frac{(x_1 + jy_1)}{(x_2 + jy_2)} . \frac{(x_2 - jy_2)}{(x_2 - jy_2)}$$

$$= \frac{x_1 x_2 + y_1 y_2}{x_2^2 + y_2^2} + j \frac{x_2 y_1 - x_1 y_2}{x_2^2 + y_2^2}$$

The complex conjugate of the complex number $z = x + jy$ is
$$z^* = x - jy = r\angle - \theta = re^{-j\theta} \qquad (2.11)$$
Thus, the complex conjugate of a complex number is found by replacing every $j$ by $-j$.
Given two complex number $z_1 = x_1 + jy_1 = r_1\angle\theta_1$ and $z_2 = x_2 + jy_2 = r_2\angle\theta_2$, their sum is
$$z_1 + z_2 = (x_1 + x_2) + j(y_1 + y_2) \qquad (2.12)$$
and their difference is
$$z_1 - z_2 = (x_1 - x_2) + j(y_1 - y_2) \qquad (2.13)$$
While it is more convenient to perform addition and subtraction of complex numbers in rectangular form, the product and quotient of two complex numbers are best done in polar or exponential form. For their product,
$$z_1 z_2 = r_1 r_2 \angle\theta_1 + \theta_2 \qquad (2.14)$$
Alternatively, using the rectangular form
$$z_1 z_2 = (x_1 + jy_1)(x_2 + jy_2)$$
$$= (x_1 x_2 - y_1 y_2) + j(x_1 y_2 + x_2 y_1)$$
$$(2.15)$$

For their quotient,

$$\frac{z_1}{z_2} = \frac{r_1}{r_2} \angle(\theta_1 - \theta_2) \tag{2.16}$$

Alternatively, using the rectangular form,

$$\frac{z_1}{z_2} = \frac{x_1 + jy_1}{x_2 + jy_2} \tag{2.17}$$

We rationalize the denominator by multiplying both the numerator and denominator by $z_2^*$.

$$\frac{z_1}{z_2} = \frac{(x_1 + jy_1)(x_2 - jy_2)}{(x_2 + jy_2)(x_2 - jy_2)}$$

$$= \frac{x_1x_2 + y_1y_2}{x_2^2 + y_2^3} + j\frac{x_2y_1 - x_1y_2}{x_2^2 + y_2^3} \tag{2.18}$$

Example 2.1

$Z_1 = 2 - 3j, Z_2 = -5 + j$

$$Z_1 + Z_2 = -3 - 2j$$

$Z_1 - Z_2 = 7 - 4j, \frac{Z_1}{Z_2} = (-1 + j)/2, Z_1.Z_2 = -7 + 17j$

## 2.2 Polar form of complex numbers

Complex number can also be expressed in terms of polar coordinates:

$$\alpha = r\cos\theta \quad , y = r\sin\theta$$

$Z = r(\cos\theta + j\sin\theta)$

Where r is the absolute value or modulus of Z and defined by

$$r = |Z| = \sqrt{x^2 + y^2} = \sqrt{\bar{Z}.Z}$$

And $\theta$ is called the argument of Z {are g(Z)}

($\theta$ will be measured in radius and positive in counter clack wise sense)

Example 2.2

$Z = 1 + j$

$r = \sqrt{1^2 + 1^2} = \sqrt{2}, \theta = tan^{-1}\frac{1}{1} = \frac{\pi}{4}, Z = \sqrt{2}(\cos\frac{\pi}{4} + j\sin\frac{\pi}{4})$

$|Z| = r = \sqrt{2}$ modulus of Z $arg(Z) = \frac{\pi}{4} \pm 2n\pi(n = 0, 1, 2...)$

The principal value of $\theta$ is $\frac{\pi}{4}$

If $Z_1 = r_1(\cos\theta_1 + j\sin\theta_1)$ and $Z_2 = r_2(\cos\theta_2 + j\sin\theta_2)$

30

$$Z_1 Z_2 = r_1 r_2 (cos(\theta_1 + \theta_2) + j\, sin(\theta_1 + \theta_2))$$

$$|Z_1 Z_2| = r_1 r_2 = |Z_1||Z_2|$$

$$arg(Z_1 Z_2) = arg(Z_1) + arg(Z_2)$$

$$\frac{Z_1}{Z_2} = \frac{r_1(cos\,\theta_1 + j\, sin\,\theta_1)}{r_2(cos\,\theta_2 + j\, sin\,\theta_2)} = \frac{r_1}{r_2}[cos(\theta_1 - \theta_2) + j\, sin(\theta_1 - \theta_2)]$$

$$\left|\frac{Z_1}{Z_2}\right| = \frac{r_1}{r_2} = \frac{|Z_1|}{|Z_2|}$$

$$arg\ \left(\frac{Z_1}{Z_2}\right) = arg(Z_1) - arg(Z_2)$$

## 2.3  Euler's formula

Euler's formula is an important result in complex variables. We derive it from the series expansion of $e^x, cos\,\theta$, and $sin\,\theta$. We know that

$$e^x = 1 + x + \frac{x^2}{2!} + \frac{x^3}{3!} + \frac{x^4}{4!} + \cdots \tag{2.19}$$

Replacing x by $j\theta$ gives

$$e^{j\theta} = 1 + j\theta - \frac{\theta^2}{2!} - j\frac{\theta^3}{3!} + \frac{\theta^4}{4!} + \cdots \tag{2.20}$$

Also,

$$cos\,\theta = 1 - \frac{\theta^2}{2!} + \frac{\theta^4}{4!} - \frac{\theta^6}{6!} + \cdots \tag{2.21a}$$

$$sin\,\theta = \theta - \frac{\theta^3}{3!} + \frac{\theta^5}{5!} - \frac{\theta^7}{7!} + \cdots \tag{2.21b}$$

so that

$$cos\,\theta + j\, sin\,\theta = 1 + j\theta - \frac{\theta^2}{2!} - j\frac{\theta^3}{3!} + \frac{\theta^4}{4!} + j\frac{\theta^5}{5!} - \cdots \tag{2.22}$$

Comparing Equations. (2.20) and (2.22), we conclude that

$$e^{j\theta} = cos\,\theta + j\, sin\,\theta \tag{2.23}$$

This is known as Euler's formula. The exponential form of representing a complex number as in Equation (2.8) is based on Euler's formula. From Equation (2.23), notice that

$$\cos\theta = Re(e^{j\theta}), \qquad \sin\theta = Im(e^{j\theta}) \tag{2.24}$$

and that

$$|e^{j\theta}| = \sqrt{\cos^2\theta + \sin^2\theta} = 1 \tag{2.25}$$

Replacing θ by -θ in Equation (2.23) gives

$$e^{-j\theta} = \cos\theta - j\sin\theta \tag{2.26}$$

Adding Equations. (2.23) and (2.26) yields

$$\cos\theta = \frac{1}{2}\left(e^{j\theta} + e^{-j\theta}\right) \tag{2.27}$$

Subtracting Equation (2.26) from Equation (2.23) yields

$$\sin\theta = \frac{1}{2j}\left(e^{j\theta} - e^{-j\theta}\right) \tag{2.28}$$

The following identities are useful in dealing with complex numbers. If $\qquad z = x + jy = r\angle\theta$, then

$$zz^* = |z|^2 = x^2 + y^2 = r^2 \tag{2.29}$$

$$\sqrt{z} = \sqrt{x + jy} = \sqrt{r}e^{j\theta/2} = \sqrt{r}\angle\theta/2 \tag{2.30}$$

$$z^n = (x + jy)^n = r^n\angle n\theta = r^n(\cos n\theta + j\sin n\theta) \tag{2.31}$$

$$z^{1/n} = (x + jy)^{1/n} = r^{1/n}\angle\theta/n + 2\pi k/$$
$$n, \qquad k = 0, 1, 2, \dots, n-1 \tag{2.32}$$

$$\ln(re^{j\theta}) = \ln r + \ln e^{j\theta} = \ln r + j\theta + j2\pi k \quad \text{(k=integer)} \tag{2.33}$$

$$e^{\pm j\pi} = -1$$
$$e^{\pm j2\pi} = 1$$
$$e^{j\pi/2} = j$$
$$e^{-j\pi/2} = -j \tag{2.34}$$

$$Re\left(e^{(\alpha+j\omega)t}\right) = Re\left(e^{\alpha t}e^{j\omega t}\right) = e^{\alpha t}\cos\omega t$$

$$Im\left(e^{(\alpha+j\omega)t}\right) = Im\left(e^{\alpha t}e^{j\omega t}\right) = e^{\alpha t}\sin\omega t \hspace{2cm} (2.35)$$

For complex number

$$Z = \alpha + jy, e^Z = e^{\alpha+jy} = e^{\alpha}e^{jy}$$

We know that $e^{jy} = \cos y + j\sin y$ ..................Euler's formula

$$e^Z = e^{\alpha}[\cos y + j\sin y]$$

Since the polar form of a complex number $Z = \alpha + jy$ is

$$Z = r(\cos\theta + j\sin\theta)$$
$$Z = re^{j\theta}$$

Note that $e^{j2\pi} = \cos 2\pi + j\sin 2\pi = 1$

We can write $w = Z^{\frac{1}{n}} = r^{\frac{1}{n}}e^{j\frac{\theta+2\pi k}{n}}$

Then, when k=n we have

$$r^{\frac{1}{n}}e^{j(\frac{\theta}{n}+2\pi)} = r^{\frac{1}{n}}e^{j\frac{\theta}{n}}e^{j2\pi} = r^{\frac{1}{n}}e^{j\frac{\theta}{n}}$$

Root at k=n is equal to the root at k=0

$$Z^{\frac{1}{n}} = r^{\frac{1}{n}}\left[\cos\frac{\theta+2\pi k}{n} + j\sin\frac{\theta+2\pi k}{n}\right]$$

Example 2.3

Solve the equation $Z^4 = -16$

Solution

$$Z = re^{j\theta} = 16e^{j\pi}$$
$$\omega_R = r^{\frac{1}{n}}e^{j\frac{\theta+2\pi k}{n}} = 16^{\frac{1}{4}}e^{j\frac{\pi+2\pi k}{4}}$$

$$k = 0, 1, 2, 3$$

$$\omega_0 = 2e^{j\frac{\pi}{4}}, \omega_1 = 2e^{j\frac{3\pi}{4}}, \omega_2 = 2e^{j\frac{5\pi}{4}}, \omega_3 = 2e^{j\frac{7\pi}{4}}$$

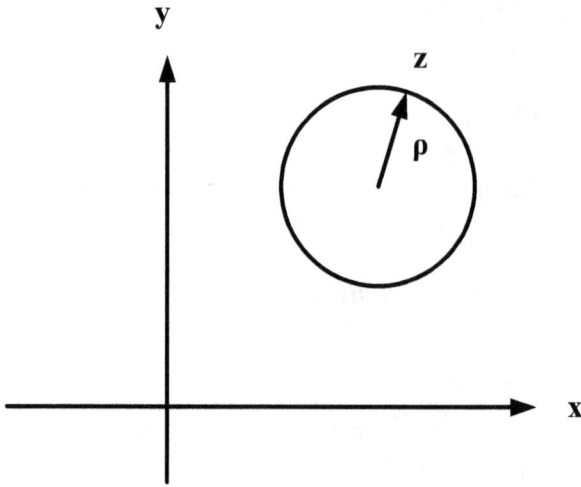

Figure 2.2 Graphical representation of a complex number of Example 2.3.

## 2.4 De Moivre's formula and it's application

If we have

$Z_1 = r_1(cos\,\theta_1 + j\,sin\,\theta_1)$

$Z_2 = r_2(cos\,\theta_2 + j\,sin\,\theta_2)$
$\vdots$
$Z_n = r_n(cos\,\theta_n + j\,sin\,\theta_n),$

$Z_1 Z_2 \dots Z_n = r_1 r_2 \dots r_n[cos(\theta_1 + \theta_2 + \cdots \theta_n) + j\,sin(\theta_1 + \theta_2 + \cdots \theta_n)]$

$or\,Z^n = r^n[cos\,n\,\theta + j\,sin\,n\,\theta]$

De Moivres formula can be used to calculus the root of $\sqrt[n]{Z}$ as follows:

$$\sqrt[n]{Z} = Z^{\frac{1}{n}} = r^{\frac{1}{n}}\left[cos\frac{\theta + 2\pi\theta k}{n} + j\,sin\frac{\theta + 2\pi\theta k}{n}\right]$$
$$\{k = 0, 1, 2, \dots\dots n - 1\}$$

Example 2.4
Solve the equation

$$Z^4 - 1 = j\sqrt{3}$$

Solution:

$$Z^4 = 1 + j\sqrt{3}, Z = (1 + j\sqrt{3})^{\frac{1}{4}}, r = \sqrt{1 + 3} = 2$$

$$\theta = tan^{-1}\sqrt{3} = \frac{\pi}{3} rad, Z = (1 + j\sqrt{3})^{\frac{1}{4}}$$

$$= 2^{\frac{1}{4}} \left[ cos \frac{\frac{\pi}{3} + 2\pi k}{4} + j sin \frac{\frac{\pi}{3} + 2\pi k}{4} \right]$$

$$k = 0, 1, 2, 3$$

The four roots are

$$k = 0 \rightarrow Z_1 + 2^{\frac{1}{4}} \left[ cos \frac{\pi}{12} + j sin \frac{\pi}{12} \right]$$

$$k = 1 \rightarrow Z_2 = 2^{\frac{1}{4}} \left[ cos \frac{2\pi}{12} + j sin \frac{2\pi}{12} \right]$$

$$k = 2 \rightarrow Z_3 = 2^{\frac{1}{4}} \left[ cos \frac{13\pi}{12} + j sin \frac{13\pi}{12} \right] = -Z_1$$

$$k = 3 \rightarrow Z_4 = 2^{\frac{1}{4}} \left[ cos \frac{19\pi}{12} + j sin \frac{19\pi}{12} \right] = -Z_2$$

Example 2.5

Find the roots of $\sqrt[3]{1}$

$$Z = 1, r = 1 \theta = tan^{-1}\frac{0}{1} = 0$$

$$\sqrt[3]{z} = 1 \left[ cos \frac{2k\pi}{3} + j sin \frac{2k\pi}{3} \right] k = 0, 1, 2$$

$$k = 0 \Rightarrow Z_1 = cos 0 + j sin 0 = 1,$$

$$k = 1 \Rightarrow Z_2 = cos \frac{\pi}{3} + j sin \frac{\pi}{3} = -\frac{1}{2} - j\frac{\sqrt{3}}{2}$$

$$k = 2 \Rightarrow Z_3 = cos \frac{4\pi}{3} + j sin \frac{4\pi}{3} = -\frac{1}{2} + j\frac{\sqrt{3}}{2}$$

$$(-\frac{1}{2} \pm j\frac{\sqrt{3}}{2})^3 = 1$$

$$(-\frac{1}{2} + j\frac{\sqrt{3}}{2}) = 1$$

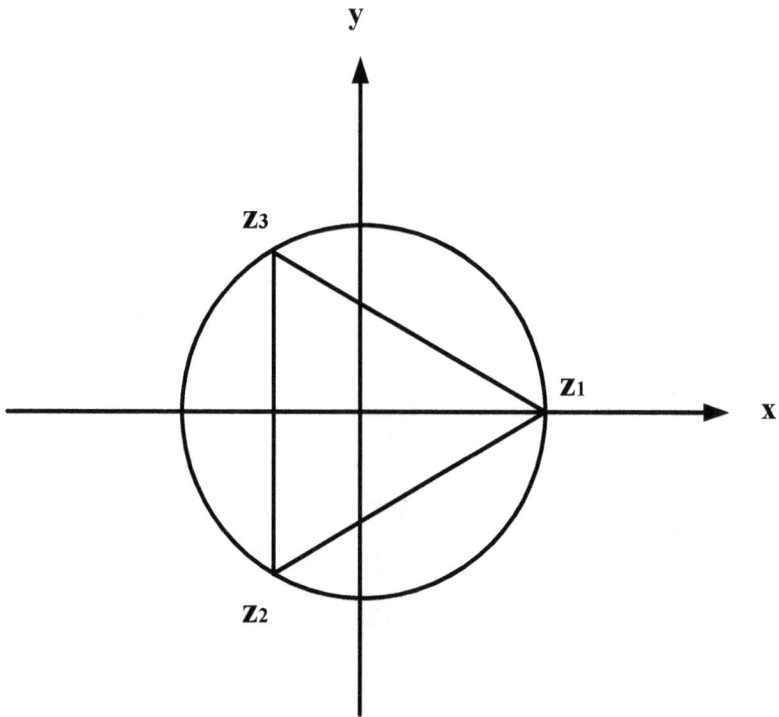

Figure 2.3 Graphical representation of a complex number.

Example 2.6

Find the roots of $(-2\sqrt{3} - 2j)^{\frac{1}{4}}$

Solution

$$Z = -2\sqrt{3} - 2j$$

$$r = 4, \theta = \frac{\pi}{6}, \sqrt[4]{Z} = 4^{\frac{1}{4}}\left[\cos\frac{\frac{\pi}{6} + 2\pi k}{4} + j\sin\frac{\frac{\pi}{6} + 2\pi k}{4}\right]$$

$$k = 0, 1, 2, 3$$

$$k = 0 Z_1 = 4^{\frac{1}{4}}\left[\cos\frac{\pi}{24} + j\sin\frac{\pi}{24}\right]$$

$$k = 1 Z_2 = 4^{\frac{1}{4}}\left[\cos\frac{13\pi}{24} + j\sin\frac{13\pi}{24}\right]$$

$$k = 2Z_3 = 4^{\frac{1}{4}}\left[\cos\frac{25\pi}{24} + j\sin\frac{25\pi}{24}\right]$$
$$k = 3Z_4 = 4^{\frac{1}{4}}\left[\cos\frac{31\pi}{24} + j\sin\frac{31\pi}{24}\right]$$

## 2.5 Complex function

Regions in complex plane

The distance between two points $z$ and $a$ is $|Z - a|$ then
$$|Z - a| = \rho$$
Represent a circle C of radius $\rho$ with center of point
The inequality $|Z - a| < \rho$

Holds for wory point inside C (interior of C) such a region is called a circular disk or open circular disk
Where $|Z - a| \leq \rho$ in the closed circular disk which consist of the interior of {c}f {c} itself further were $\rho_1 < |Z - a| \leq \rho_2$

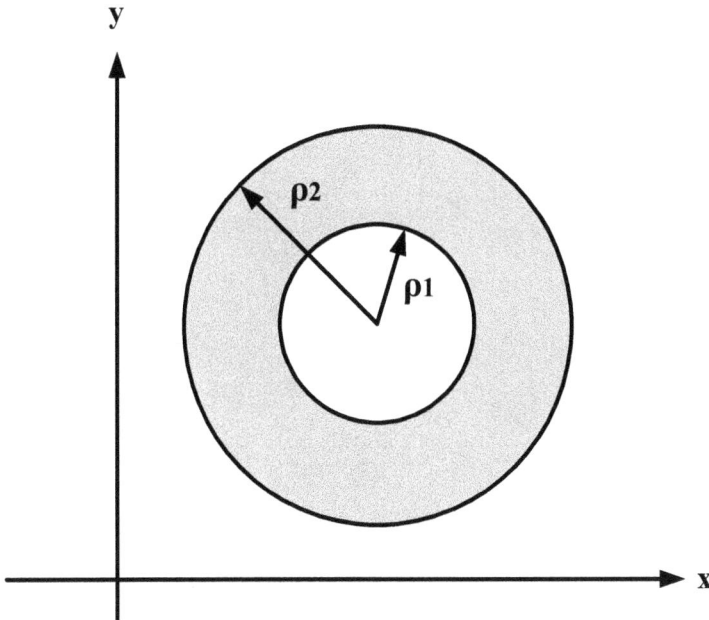

Figure 2.4 Regions in complex plane

Represent the region between two concenter circles of radii $\rho_1$ and $\rho_2$ ($>\rho_1$) which is also called circular ring or open annulus.
The equation $|Z| = 1$ represent a unit circle radius $=1$ center of origin.

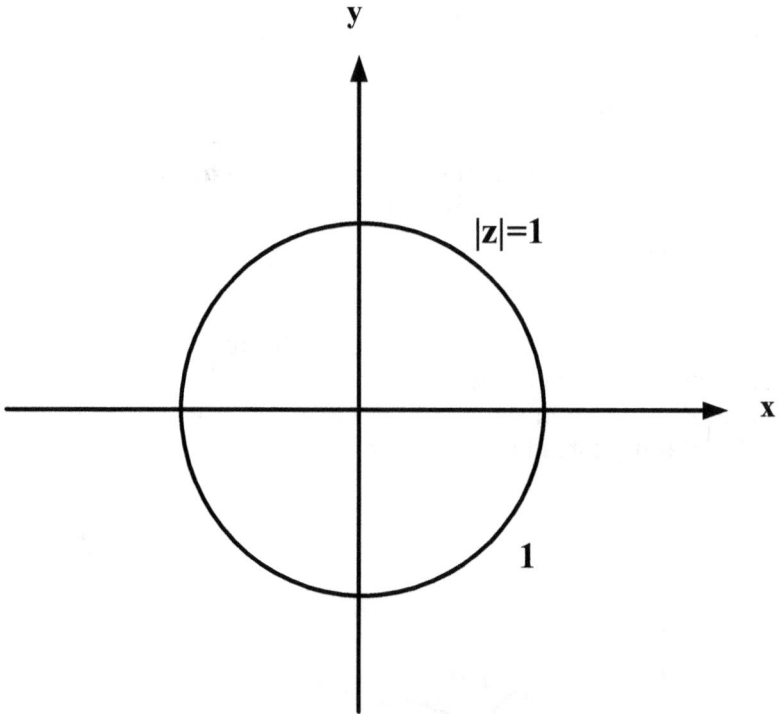

Figure 2.5 Region in complex plane with $|z| = 1$.

Example 2.7

Describe the set of points
1-      $|Z - 2| = 1$
2-      $1 \le |Z - 3| \le 2$
Solution

Closed circular disk of radius 1 with center at
a.      $Z = 2 + j2$
b.      Closed annulus with center at $Z = 3 + j3$ and radii $\rho_1 = 1$ and $\rho_2 = 2$

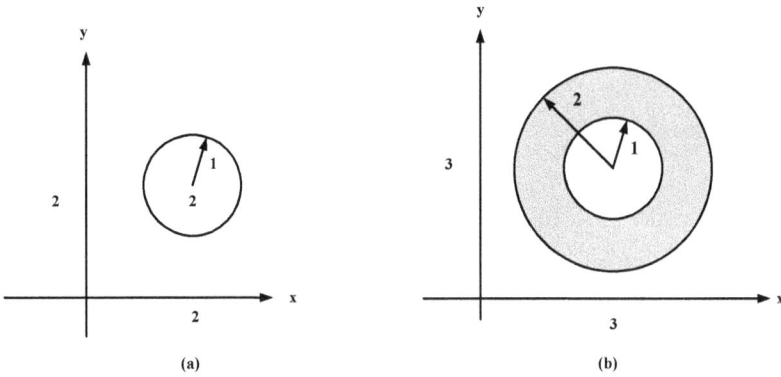

Figure 2.6 Graphical representation of a complex number of Example 2.7.

Definition of complex function

If for a set S of complex number, there is a rule which assigns to each Z in **S** a unique complex number $\omega$, then

$\omega = f(Z)$ or $\omega = g(Z)$ or $\omega(Z)$

Z various in **S** and it is called complex variable

The set of complex numbers, which $\omega$ assumes is called the ranges .

S is called the domain of the complex function

Let u be the real part of $\omega$ and v be the imaginary part of $\omega$then

$$\omega = u + jv = f(x, y)$$
$$\omega = u(x, y) + jv(x, y)$$

As example

let $\omega = f(Z) = Z^2 + 3Z$ therefor

$$(x + jy)^2 + 3(x + jy)$$
$$= \underbrace{x^2 - y^2 + 3x}_{u} + j \underbrace{(2xy + 3y)}_{v}$$

Exercise:  Find the Real and Imaginary parts of $\omega = f(Z) = \frac{1}{1-Z}$

2.6 Limits, Continuity, Derivative

We have $\omega = f(z)$ is said to be single value function of **z**. if only one value of $\omega$corresponds to each value of **z**

39

Example 2.8

$$\omega = Z^2, \omega = Z^{\frac{1}{2}} \rightarrow$$
$$\downarrow$$

where $Z^{\frac{1}{2}}$ is a multiple value function (to each value of $z$ there are two Value of $\omega$)
and $\omega$ is the single value Function.

If $\omega(z)$ is a single valued function defined all points in the reih bow-head of some point $z$ then the statement of the limit of $\omega(z)$ as $z \rightarrow z_0$ is

$$\lim_{z \to z_0} \omega(z) = \lim_{z \to z_0} f(z) = \omega_0$$

Theorem of limits

If $\lim_{z \to z_0} \omega(z) = \omega_0$ and $\lim_{z \to z_0} G(z) = G_0$ then :

1- $\lim_{z \to z_0} \{\omega(z) \pm G(z)\} = \lim_{z \to z_0} \omega(z) \pm \lim_{z \to z_0} G(z) = \omega_0 \pm G_0$

2- $\lim_{z \to z_0} (\omega(z).G(z)) = \left\{\lim_{z \to z_0} \omega(Z)\right\}\left\{\lim_{z \to z_0} G(Z)\right\} = \omega_0.G_0$

3- $\lim_{z \to z_0} \dfrac{\omega(z)}{G(z)} = \dfrac{\lim_{z \to z_0} \omega(z)}{\lim_{z \to z_0} G(z)} = \dfrac{\omega_0}{G_0} = G_0 \neq 0$

A function $f(z)$ is said to be continuous at points $z = z_0$ if
1- $f(z_0)$ is defined
2- $\lim_{z \to z_0} f(z) = f(z_0)$
A function $f(z)$ is said to be differential at point $z = z_0$ if
$\lim_{z \to z_0} \dfrac{f(z_0 + \Delta z) - f(z_0)}{\Delta z} = f'(z)$ exist
This limit is called the derivative of $f(z)$ at point $z_0$

Example 2.9

Show that the function $f(z) = z^2$ is differential and has $f'(z) = 2z$

40

Solution

$$\lim_{\Delta z \to 0} \frac{(z + \Delta z)^2 - z^2}{\Delta z} = \lim_{\Delta z \to 0} \frac{z^2 + 2z\Delta z + \Delta z^2 - z^2}{\Delta z}$$

$$\lim_{\Delta z \to 0} \to (2z + \Delta z) = 2z$$

Moreover, if $\omega_1 = f_1(z)$ and $\omega_2 = f_2(z)$ are differentiable functions then

1- $\frac{\partial}{\partial z}(\omega_1 \pm \omega_2) = \frac{\partial \omega_1}{\partial z} \pm \frac{\partial \omega_2}{\partial z}$

2- $\frac{\partial}{\partial z}(\omega_1 . \omega_2) = \omega_1 \frac{\partial \omega_2}{\partial z} + \omega_2 \frac{\partial \omega_1}{\partial z}$

3- $\frac{\partial}{\partial z}\left(\frac{\omega_1}{\omega_2}\right) = \frac{\omega_2 \frac{\partial \omega_1}{\partial z} - \omega_1 \frac{\partial \omega_2}{\partial z}}{\omega_2^2}$

4- $\frac{\partial}{\partial z}(\omega^n) = n\omega^{n-1} \frac{d\omega}{dz}$

## 2.7 Cauchy–Riemann equation

$$\omega = f(z) = u(\alpha, y) + jv(\alpha, y)$$

Suppose $f(z)$ to be defined and continuous in same neighbor had of an arbitrary fixed point and differentiable of z so that

$$f'(z) = \lim_{\Delta z \to 0} \frac{f(z + \Delta z) - f(z)}{\Delta z}$$

At that point exists.

41

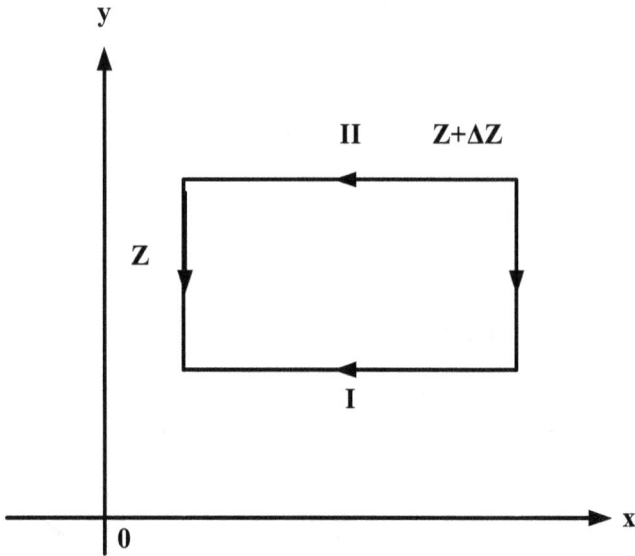

Figure 2.7 Region of Cauchy–Riemann Equation.

$\Delta z = \Delta \alpha + j\Delta y$ and may approach to zero along any path

Let $\Delta z \to 0$ along path I

(i.e $\Delta y \to 0$ first, then $\Delta \alpha \to 0$)

Then after $\Delta y \to 0$ then $\Delta z = \Delta \alpha$ and

$$f'(z) = \lim_{\Delta x \to 0} \frac{[u(x + \Delta x, y) + jv(x + \Delta x, y)] - [u(x, y) + jv(x, y)]}{\Delta x}$$

$$= \lim_{\Delta x \to 0} \frac{u(x + \Delta x, y) - u(x, y)}{\Delta x} + j \lim_{\Delta x \to 0} \frac{v(x + \Delta x, y) - v(x, y)}{\Delta x}$$

$$f'(z) = \frac{\partial u}{\partial x} + j \frac{\partial v}{\partial x} \tag{2.37}$$

Let $\Delta z \to 0$ along path I ($\Delta x \to 0$, first , then $\Delta y \to 0$) after $\Delta x \to 0, \Delta z = j\Delta y$. So we obtain:

$$f'(z) = \lim_{\Delta y \to 0} \frac{u(x, y+\Delta y) - jv(x, y+\Delta y) - [u(x, y) + jv(x, y)]}{j\Delta y}$$

$$= \lim_{\Delta y \to 0} \frac{u(x, y+\Delta y) - u(\alpha, y)}{j\Delta y} + j \lim_{\Delta y \to 0} \frac{v(x, y+\Delta y) - v(x, y)}{j\Delta y} \tag{2.38}$$

$$= \frac{1}{j} = -j$$

42

$$f'(z) = -j\frac{\partial u}{\partial y} + \frac{\partial v}{\partial y}$$

More important by equating the real and imaginary part set Equations 2.37 and 2.38 where

$$\frac{\partial u}{\partial \alpha} = \frac{\partial v}{\partial y}, \frac{\partial u}{\partial y} = -\frac{\partial v}{\partial \alpha}$$    It's called Cauchy-Riemann equations

Example 2.10

If $f(z) = z^2$    find $f'(z)$

Solution

$$f(z) = z^2 = (\alpha + jy)^2 = \underbrace{\alpha^2 - y^2}_{u} + j\underbrace{2\alpha y}_{v}$$

$$u = \alpha^2 - y^2, v = 2\alpha y$$

$$\frac{\partial u}{\partial \alpha} = 2\alpha, \frac{\partial v}{\partial \alpha} = 2y$$

$$f'(z) = \frac{\partial u}{\partial \alpha} + j\frac{\partial v}{\partial \alpha} = 2\alpha + j2y = 2(\alpha + jy) = 2z$$

or $f'(z) = \frac{\partial v}{\partial y} - j\frac{\partial u}{\partial y} = 2\alpha + j2y = 2(\alpha + jy) = 2z$

## 2.8 Analytic function

Definition: a function $f(z)$ is said to be analytic in domain Diff $f(z)$ is defined and differentiable at all points of D

$$\omega = f(z) = u(\alpha, y) + jv(\alpha, y)$$

If $f(z)$ is analytic in domain D, then u and v satisfy then so called Cauchy Riemann equation everywhere in D. $\frac{\partial v}{\partial x} = \frac{\partial v}{\partial y}, \frac{\partial v}{\partial y} = -\frac{\partial v}{\partial x}$

$$\frac{\partial u}{\partial \alpha} = \frac{\partial u}{\partial y}, \frac{\partial u}{\partial y} = -\frac{\partial u}{\partial \alpha}$$

Ex : for $f(z) = z^2 = (\alpha + jy)^2 = \alpha^2 - y^2 + j2\alpha y$

$$u = \alpha^2 - y^2, v = 2\alpha y$$

$\frac{\partial u}{\partial \alpha} = 2\alpha, \frac{\partial v}{\partial y} = 2\alpha, \frac{\partial u}{\partial y} = -2y, \frac{\partial v}{\partial \alpha} = 2y$

Then Cauchy Riemann equation are satisfying here the derivative $f'(z)$ exists at all points it the z-plane $\Rightarrow f(z) = z^2$ is analytic

Example 2.11
$$f(z) = z\bar{z} = (\alpha + jy)(\alpha - jy) = \underbrace{\alpha^2 + y^2}_{u} + \underbrace{0}_{v}$$
$$u = \alpha^2 + y^2, v = 0$$

$$\frac{\partial u}{\partial \alpha} = 2\alpha, \frac{\partial u}{\partial y} = 2y, \frac{\partial v}{\partial \alpha} = 0, \frac{\partial v}{\partial y} = 0$$

Cauchy Riemann equation are $2\alpha = 0, 2y = 0$
Which are satisfied only $\alpha = 0$
$y = 0$ ( i.e $z = 0$ at origin ) therefore $f'(z)$ exists at $z = 0$ only or $f(z) = z\bar{z}$ is not analytic

Exercise: Is $f(z) = \bar{z}$ analytic function?
Theorem (1) : if u and v real single valued functions $\alpha$ and y which with their first partial derivatives are continuous the rough on region D, then the Cauchy Riemann equation
$$\frac{\partial u}{\partial \alpha} = \frac{\partial v}{\partial y}, \frac{\partial u}{\partial y} = -\frac{\partial v}{\partial \alpha}$$

Are both necessary and sufficient conditions that $f(z) = u(\alpha, y) + jv(\alpha, y)$ be analytic in D
Diff the Cauchy Riemann equations :
$$\frac{\partial^2 u}{\partial \alpha \partial y} = \frac{\partial^2 v}{\partial y^2}, \frac{\partial^2 u}{\partial y \partial \alpha} = -\frac{\partial^2 v}{\partial \alpha^2}$$

$$\frac{\partial^2 u}{\partial \alpha \partial y} = \frac{\partial^2 v}{\partial y \partial \alpha}, \frac{\partial^2 v}{\partial \alpha^2} = -\frac{\partial^2 v}{\partial y^2}$$

$$\nabla^2 v = \frac{\partial^2 v}{\partial \alpha^2} + \frac{\partial^2 v}{\partial y^2} = 0$$

Similarly, we can prove that
$$\frac{\partial^2 u}{\partial \alpha^2} + \frac{\partial^2 u}{\partial y^2} = 0 = \nabla^2 u$$

Theorem (2)
The real part and imaginary part of a complex function $f(z) = u(\alpha, y) + jv(\alpha, y)$, that is analytic in domain D are solutions of Laplace equations
$$\nabla^2 u = \frac{\partial^2 u}{\partial \alpha^2} + \frac{\partial^2 u}{\partial y^2} = 0, \nabla^2 v = \frac{\partial^2 v}{\partial \alpha^2} + \frac{\partial^2 v}{\partial y^2} = 0$$

In D and have continuous partial derivatives in D

$u(\alpha, y) + v(\alpha, y)$ are called harmonic function and $v(\alpha, y)$ is the conjugate harmonic functions

Example 2.12
 Show that
$f(z) = e^z$ is analytic and find $f'(z)$

Solution

$$f(z) = e^z = e^{a+iy} = e^a . e^{jy} = e^a (\cos y + j \sin y)$$

$$f(z) = \underbrace{e^a \cos y}_{u} + j \underbrace{e^a \sin y}_{v}$$

$$\frac{\partial u}{\partial \alpha} = e^a \cos y, \frac{\partial u}{\partial y} = -e^a \sin y$$

$$\frac{\partial v}{\partial \alpha} = e^a \sin y, \frac{\partial v}{\partial y} = e^a \cos y$$

Cauchy Riemann equations states that
$$\frac{\partial u}{\partial \alpha} = \frac{\partial v}{\partial y} = e^a \cos y$$
$$\frac{\partial u}{\partial y} = -\frac{\partial v}{\partial \alpha} = -e^a \sin y$$

$f(z) = e^z$ is analytic
Now to find $f'(z)$
$$f'(z) = \frac{\partial u}{\partial \alpha} + j \frac{\partial v}{\partial \alpha} = e^a \cos y + j e^a \sin y$$

Exercise: Show that $f(z) = e^{\overline{z}}$ is nowhere analytic .

Example 2.13

Find the most general analytic function $f(z) = u(\alpha, y) + jv(\alpha, y)$ given that $xy = u$

Solution
$f(2)$ analytic $\Rightarrow \dfrac{\partial u}{\partial \alpha} = \dfrac{\partial v}{\partial y}$

And $\dfrac{\partial u}{\partial y} = -\dfrac{\partial v}{\partial \alpha}$      C.R equations

45

$$\frac{\partial u}{\partial x} = y = \frac{\partial v}{\partial y} \rightarrow v(\alpha, y) = \frac{1}{2}y^2 + h(x)$$

$$\frac{\partial v}{\partial \alpha} = \frac{\partial h(x)}{\partial \alpha} = -\frac{\partial u}{\partial y} = -x$$

$$h(\alpha) = -\frac{1}{2}\alpha^2 + c$$

$$v(x, y) = \frac{1}{2}y^2 - \frac{1}{2}\alpha^2 + c$$

$$\omega = u + jv = \alpha y + j(\frac{1}{2}y^2 - \frac{1}{2}\alpha^2 + c)$$

$$2j\omega = 2j\alpha y - y^2 + \alpha^2 - 2c$$

$$2j\omega = z^2 - 2c$$

$$\omega = -j\frac{z^2}{2} + jc$$

Exercise: Find the analytic function $f(z) = u + jv$ if $v = \alpha y$

Example 2.14

Find the conjugate the harmonic function of the harmonic function $u = \alpha^2 - y^2$

Solution: We have

$$\frac{\partial u}{\partial \alpha} = 2\alpha, \frac{\partial u}{\partial y} = -2y$$

The conjugate of $u(\alpha, y)$ or $(v((\alpha, y))$ must satisfy the C.R equations

$$\frac{\partial u}{\partial \alpha} = -\frac{\partial v}{\partial y} = 2\alpha, \frac{\partial u}{\partial y} = -\frac{\partial v}{\partial \alpha} = -2y$$

$$\frac{\partial v}{\partial y} = 2\alpha \rightarrow v = 2\alpha y + h(x)$$

$$\frac{\partial u}{\partial \alpha} = 2y + \frac{\partial h(x)}{\partial \alpha}, \frac{\partial u}{\partial \alpha} = 2y \Rightarrow \frac{\partial h(x)}{\partial \alpha} = 0 \rightarrow h(x) = 0$$

$$v = 2\alpha y + c$$

The analytic function

$$f(z) = a^2 - y^2 + j(2\alpha y + c) = z^2 + jc$$

Trigonometric an by parabolic function
we have

$$\cos\alpha = \frac{1}{2}(e^{j\alpha} + e^{-j\alpha})$$

$$\sin\alpha = \frac{1}{2j}(e^{j\alpha} - e^{-j\alpha}) \qquad \text{Euler's formula } \alpha \text{ is real}$$

For complex variables

$$\cos z = \frac{1}{2}(e^{jz} + e^{-jz}), \cosh z = \frac{e^z + e^{-z}}{2}$$

$$\sin z = \frac{1}{2j}(e^{jz} - e^{-jz}), \sinh z = \frac{e^z - e^{-z}}{2}$$

For the more

$$\tan z = \frac{\sin z}{\cos z}, sccz = \frac{1}{\cos z}$$

$$csc = \frac{1}{\sin z} \quad ,e^z \text{ is analytic}$$

sin z and cos z are analytic
Note that $\cos(\alpha + jy)$

$$\cos z = \cos\alpha \cosh y - j \sin\alpha \sinh y$$

$$\sin z = \sin\alpha \cosh y + j \cos\alpha \sinh y$$

$$\sin jz = j \sinh z$$

$$\sinh jz = j \sin z$$

$$\cos jz = \cosh z, \cosh jz = \cos z$$

## 2.9 Complex integral

2.9.1 Line Integral in complex plane

Let C be a smooth curve in the complex z plane and may be presented by $Z(t) = \alpha(t) + jy(t)$ $\qquad$ (2.39)
Let $f(z)$ be continuous function, which is defined (at least) at each point of C. Then the line integral of $f(z)$ along is defined:
$\int_c f(z).dz$ $\qquad$ (2.40)
C is the path of integration
If $f(z) = u(\alpha, y) + jv(\alpha, y)$ then
$\int_c f(z).dz = \int_c (u + jv)(\partial\alpha + j\partial y)$
$= \int_c u\partial x - \int_c v\partial y + j[\int_c u\partial y + \int_c v\partial x]$ $\qquad$ (2.41)
If c is defined by Equation 2.39, where the positive direction along C corresponds to the sense of increasing value of the parameter, then the integration

$$\int_c f(z)dz = \int_a^b (u + jw)(\alpha + jy)dt$$

$$= \int_a^b f(z,t)\dot{Z}(t)dt \qquad (2.42)$$

Example 2.15

Represent the following cover in the form $z = z(t)$
a- line segment from $\quad z = 0$ to $z = 3 - 12j$
b -line segment from $z_1 = 1 - j$ to $\quad z_2 = 9 - 5j$
c- $|z| = 2|z - 0| = z$
d-$|z - j| = 2$
e-$|z + 1| = 3$
f- $y = \alpha^2 \qquad$ from $(0,0)$ to $(2,4)$
g-$|z - (2 - 3j)| = 6$
h- $\alpha^2 + 4y^2 = 4$
i- $4\alpha^2 + y^2 = 4$
j-$4(\alpha - 1)^2 + y^2 = 4$

Solution:

a- line segment from $\quad z = 0$ to $z = 3 - 12j$
$$z(t) = 0 + 3t - 12j$$

$$\text{or} \quad = (3 - j12)t \qquad\qquad 0 \le t \le 1$$
$$= (1 - j4)t \qquad\qquad 0 \le t \le 3$$

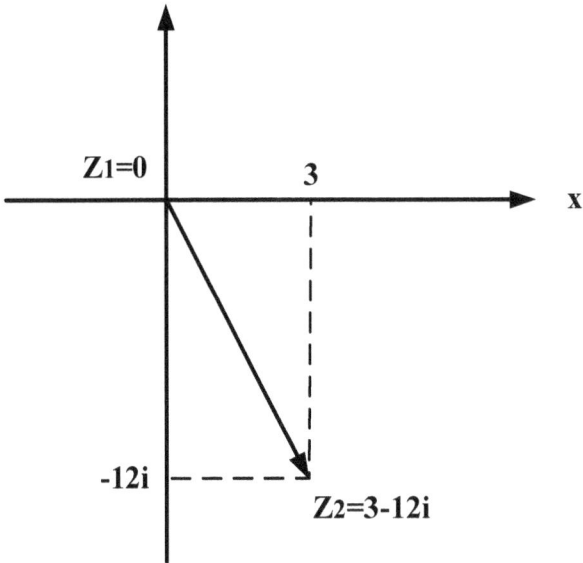

Figure 2.8 Graphical representation of a complex number of Example 2.15 (a)

b -line segment from $z_1 = 1 - j$ to $\quad z_2 = 9 - 5j$
$$z(t) = 1 - j + (8 - 4j)t \qquad\qquad 0 \le t \le 1$$
Or $z(t) = 1 - j + (2 - j)t \qquad\qquad 0 \le t \le 4$

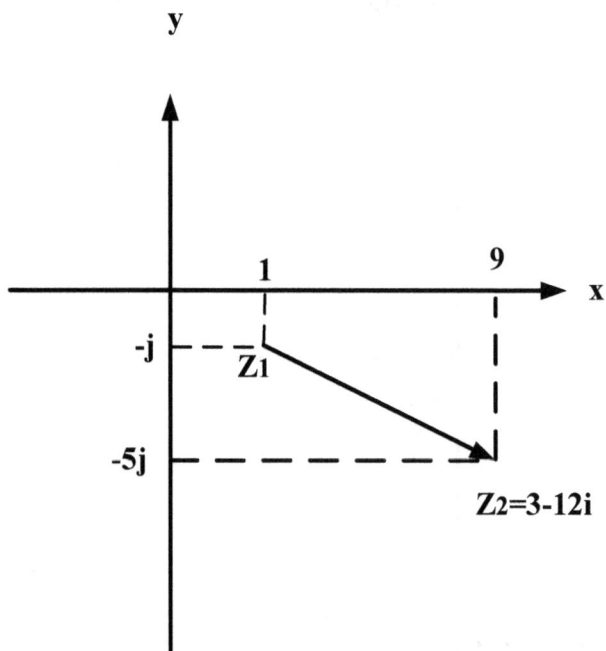

Figure 2.9 Graphical representation of a complex number of Example 2.15 (b)

c- $|z| = 2|z - 0| = z$

$$z(t) = 2\cos t + j2 \sin t$$

$z(t) = 2e^{jt} \, 0 \leq t \leq 2\pi$

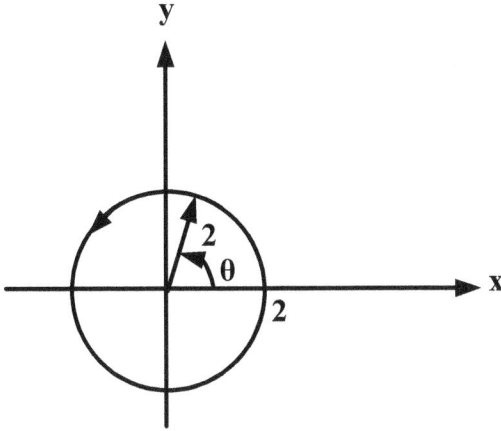

Figure 2.10 Graphical representation of a complex number of Example 2.15 (c).

d-$|z - j| = 2$

$$z(t) = j + 2e^{jt} \qquad 0 \leq t \leq 2\pi$$

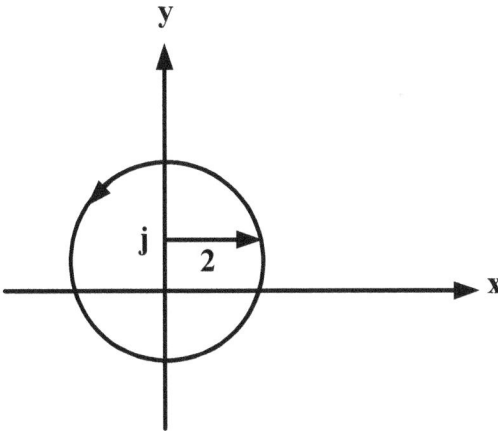

Figure 2.11 Graphical representation of a complex number of Example 2.15 (d).

e-

$$|z + 1| = 3$$

$$z(t) = 1 + 3e^{jt} \qquad\qquad 0 \le t \le 2\pi$$

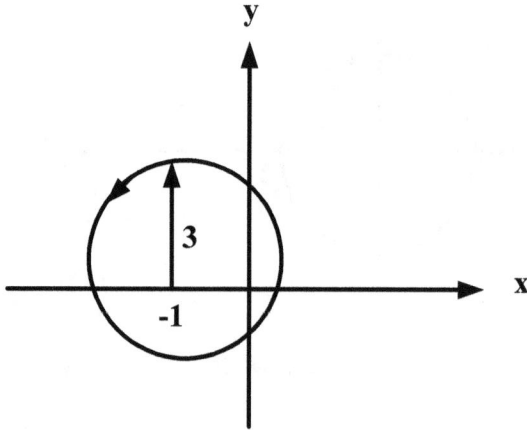

Figure 2.12 Graphical representation of a complex number of Example 2.15 (e).

f-  $\quad y = \alpha^2 \qquad$ from $(0, 0)$ to $(2, 4)$
$z(t) = t + jt^2 \qquad\qquad 0 \le t \le 2$

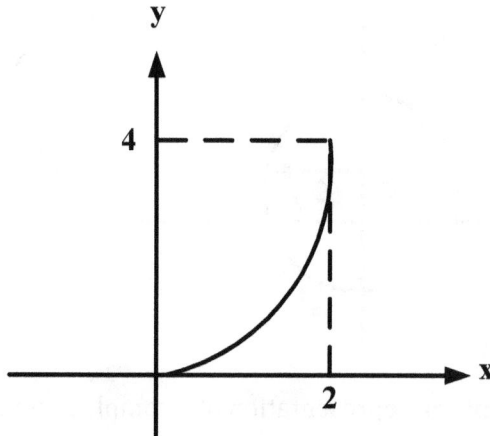

Figure 1.13 Graphical representation of a complex number of Example 2.15 (f).

g-
$$|z - (2 - 3j)| = 6$$
$$z(t) = 2 - 3j + 6e^{jt} \qquad 0 \le t \le 2\pi$$

h- $a^2 + 4y^2 = 4$

$\dfrac{a^2}{4} + \dfrac{y^2}{1} = 1$  ellipse

$$z(t) = 2cost + j\,sint \qquad 0 \le t \le 2\pi$$

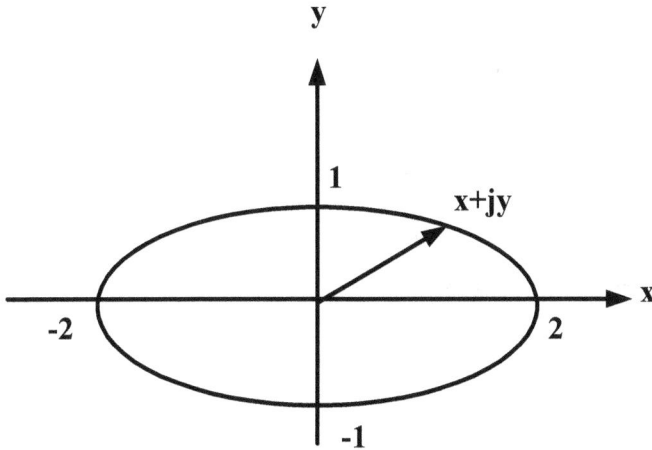

Figure 2.14 Graphical representation of a complex number of Example 2.15 (h).

i- $4a^2 + y^2 = 4$

$\dfrac{a^2}{1} + \dfrac{y^2}{4} = 1$

$$z(t) = cos\,t + 2j\,sin\,t \quad 0 \le t \le 2\pi$$

$$z(t) = a\,cos\,t + b\,sin\,t$$

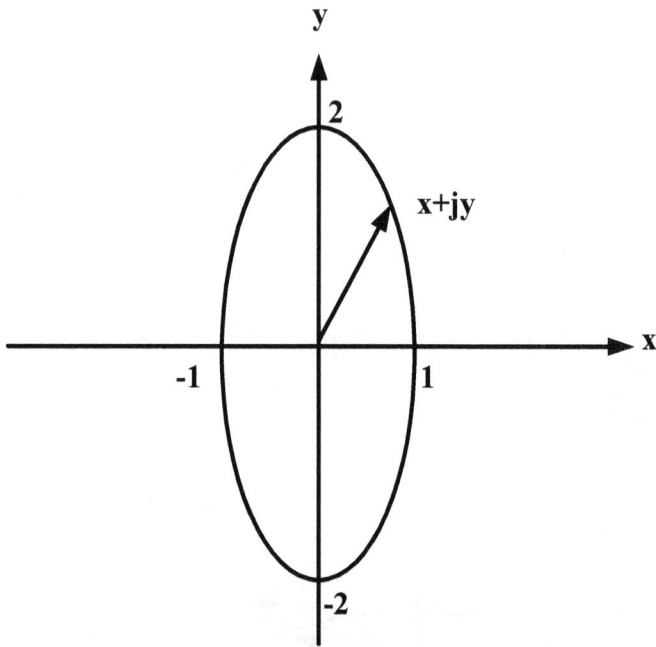

Figure 2.15 Graphical representation of a complex number of Example 2.15 (i)

j-4$(\alpha - 1)^2 + y^2 = 4$

$$\frac{(\alpha - 1)^2}{1} + \frac{y^2}{4} = 1$$
$$z(t) = 1 + \cos t + j2 \sin t \quad 0 \le t \le 2\pi$$

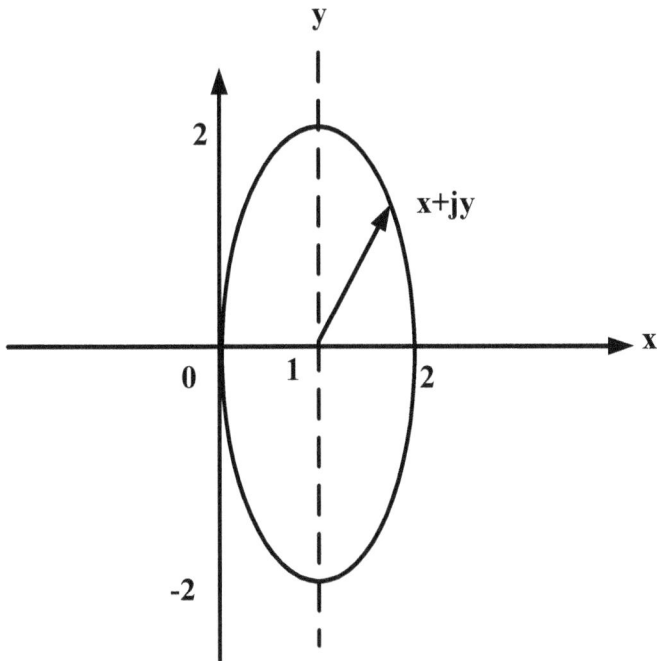

Figure 2.16 Graphical representation of a complex number of Example 2.15 (j).

Example 2.16

Integrate $z^2$ along the line segment from zero to $2 + j$.

Solution

$\int_c z^2 \partial z$

$\qquad z(t) = 0 + (2 + j)t \qquad 0 \le t \le 1$

$\qquad \dfrac{dz}{dt} = 2 + j$

$\qquad dz = (2 + j)\partial t = \int_c z^2 \partial z = \int_0^1 [(2 + j)t]^2 \ (z + j)\partial t$

$$= (2+j) \int_0^1 (2t + jt)^2 \, \partial t = (2+j) \int_0^1 (4t^2 - t^2 + 4t) \partial t$$

$$= (2+j) \int_0^1 (3t^2 + j4t^2) \partial t = (2+j) \left[ t^3 + j\frac{4}{3}t^3 \right]_0^1$$

$$= (2+j)(1 + j\frac{4}{3})$$

Exercise: Integrate $z^2$ along the line segment from $-3 + j2$ to $-4 + 5j$

Example 2.17: (Integration of non analytic function)
Integrate $f(z) = $ Real $Z = x$ along
a) $c_1$: the line segment from 0 to $1 + j$
b) $c_2$: the real axis from 0 to 1 and vertically to $1 + j$

Solution
along $c_1$

$$z(t) = (1 + j)t \quad 0 \le t \le 1$$

$$\frac{\partial z(t)}{\partial t} = (1 + j)$$

$$\partial z(t) = (1 + j)\partial t, \int_c Real\ z \ \partial z = \int_0^1 (1 + j)\partial t = \frac{1}{2}(1 + j)$$

Along $c_2$
The Real axis

$$z(t) = t \qquad\qquad 0 \le t \le 1$$

$$\partial z = \partial t$$
Vertical part

$$z(t) = 1 + jt \quad 0 \le t \le 1$$

$$\partial z = j\partial t$$

$$\int_{c_2} Re\ a\ lz.\partial z = \int_0^1 t\partial t + \int_0^1 j\partial t = \frac{1}{2} + j$$

56

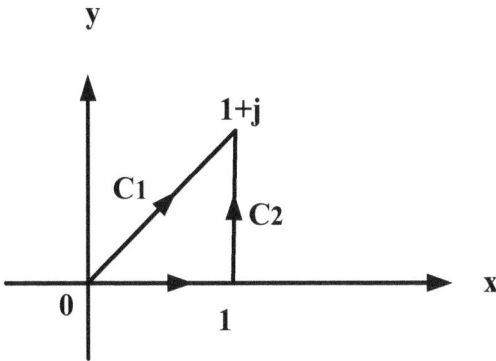

Figure 2.17 Graphical representation of a complex number of Example 2.17.

Exercise: Find $\int z\partial z$ along $c_1$ the line segment from 0 to $3 + 2j$ , $c_2$ from 0 vertically to $2j$ and then horizontally to $3 + 2j$

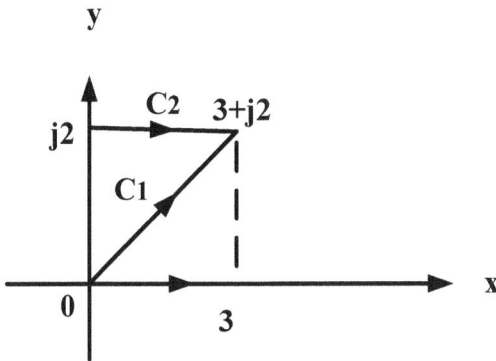

Figure 2.18 Graphical representation of a complex number of Exercise above.

Example 2.18

Evaluate $\int_c (z + 2)^{-1}\partial z$ where $c: |z - 2| = 3$
$$z(t) = 2 + 3e^{jt} \qquad 0 \le t \le 2\pi$$

Solution

$$\partial z = 3je^{jt}\partial t$$

$$\int_c (z-2)^{-1}\partial z = \int_0^{2\pi} (3\,e^{jt})^{-1}3je^{jt}\partial t = \int_0^{2\pi} j\partial t = 2\pi j$$

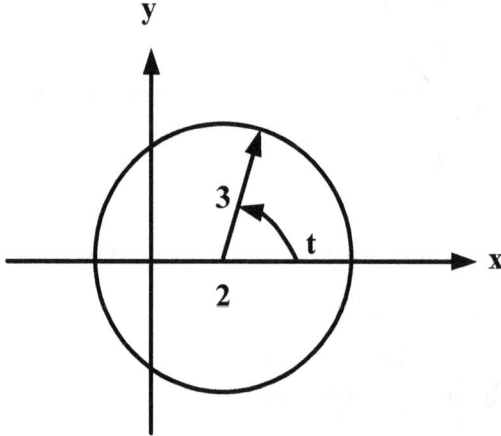

Figure 2.19 Graphical representation of a complex number of Example 2.18.

This integral along counter clockwise direction. If this required .along clock wise direction it will be $-2\pi j$ because $\int_{2\pi}^0$

Note Remark
If $f(z) = (z-z)^m$ where $m$ is integer and $z_0$ is constant, then the line integer in the counter-clockwise around the circle C of radius $\rho$ with center at $z_0$ is

$$z(t) = z_0 + \rho e^{jt} \quad 0 \le t \le 2\pi$$

$$\partial z = \rho je^{jt}\partial t$$

$$z - z_0 = \rho e^{jt}$$

Therefore, we obtained $\int_c (z-z)^m \partial z = \int_0^{2\pi} \rho^m je^{jmt}\rho je^{jt}\partial t$

$$= j\rho^{m+1} \int_0^{2\pi} e^{j(m+1)t}\partial t$$

$$= j\rho^{m+1} \left[ \int_0^{2\pi} \{cos(m+1)t + j\,sin(m+1)t\partial t\} \right]$$

$$= 2\pi jatm = -1$$

$$= 0atm \neq -1$$

## 2.9.2 Basic properties of line integrals

1- $\int_c f(z)\partial z = \int_{c1} f(z)\partial z + \int_{c2} f(z)\partial z$

2- $\int_{z_0}^{z} f(z)\partial z = -\int_{z}^{z_0} f(z)\partial z$

3- $\int_c [k_1 f_1(z) + k_2 f_2(z)\partial z] = k_1 \int_c f_1(z)\partial z + k_2 \int_c f_2(z)\partial z$

Example 2.19

Evaluate $\int_c f(z)dzwheref(z) = z^4 - z^2 - z^{-1}$
$c_i$unit circle (counter clack wise)

Solution

$$\int_c f(z)\partial z = \int_c z^4 \partial z - \int_c z^2 \partial z + \int_c z^{-1} \partial z$$
$$= 2\pi j$$

Example 2.20

Evaluate $\int_c f(z)\partial z$where $f(z) = z^{-5} + z^3$ and c the low arc of the unit circle from +1 to-1

Solution

$$z(t) = z_0 + \rho e^{jt}$$

$$z(t) = e^{jt} 0 \leq t \leq -\pi$$

$$\partial z = j e^{jt} \partial t$$

$$\int_c f(z)\partial z = \int_0^{-\pi} \left\{ [e^{jt}]^{-5} + [e^{jt}]^3 \right\} j e^{jt} \partial t$$

$$= j \int_0^{\pi} (e^{-4tj} + e^{j4t})\partial t = 2j \int_0^{-5} \cos 4t\partial t = 0$$

59

Example 2.21

Evaluate $\int_c f(z)\partial z$ where $f(z) = e^z$ and c is the line segment from 1 to $1 + j\frac{\pi}{2}$

Solution

$$z(t) = 1 + j\frac{\pi}{2}t \qquad 0 \le t \le 1$$

$$\partial z = j\frac{\pi}{2}\partial t , \qquad \int_c f(z)\partial z = \int_0^1 e^{1+j\frac{\pi}{2}t} . j\frac{\pi}{2}\partial t$$

$$= j\frac{\pi}{2}\left[\frac{e^{1+j\frac{\pi}{2}t}}{j\frac{\pi}{2}}\right]_0^1 = e^{1+j\frac{\pi}{2}} - e^1 = e^1(e^{j\frac{\pi}{2}} - 1)$$

$$e\left\{\cos\frac{\pi}{2} + j\sin\frac{\pi}{2} - 1\right\} = e[j-1]$$

## 2.10 Cauchy's integral theorem

If $f(z)$ is analytic in a simply connected bounded domain D. Then for every simply closed path C in D
$\int_c f(z)\partial z = 0$........1 (proof see why lie page 902 )
$D_1: |z| \le \rho$  bounded
$D_2: |z| > \rho$  unbounded

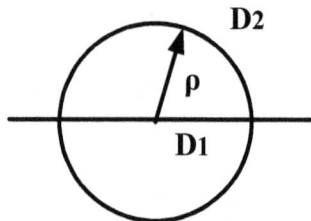

Figure 2.20 Graphical representation of a Cauchy's Integral boundaries.

Note: $\int_c e^z \partial z = 0$ for any closed path because $e^z$ is analytic for all $z$.

Example 2.22

$\int_C \frac{1}{z^2} \partial z = \int_C z^{-2} \partial z$ Where C is the unit circuit here $\frac{1}{z^2}$ in circuit analytic at $z = 0$ but we see that this integrant $=0$, therefore the condition that $f(z)$ be analytic in D is sufficient rather than necessary for Cauchy's theorem to be true

Independent of path

$$\int_C f(z)\partial z = \int_{C_1} f(z)\partial z + \int_{C_2^*} f(z)\partial z = 0$$

$$\int_{C_2^*} f(z)\partial z = -\int_{C_2} f(z)\partial z$$

$$\int_{C_2} f(z)\partial z = \int_{C_1} -f(z)\partial z$$

Note: This is holds for any paths entirety in the simply connected domain D, where $f(z)$ is analytic and joining any points $z_1$ and $z_2$ in D

Note that : $\int_C (z - z_0)^m \partial z = \begin{cases} 2\pi j & m = -1 \\ 0 & m \neq -1 \text{ and integer} \end{cases}$

Where c is any counter containing the path $z$ .in it's anterior and the integration is taken around c in counter clock wise direction

Example 2.23

Evaluate the integral $\int_C \frac{z^2 - \frac{1}{3}}{z^3 - z} \partial z$ where $c: \left|z - \frac{1}{2}\right| = 1$ (counter clockwise)

Solution

$$f(z) = \frac{z^2 - \frac{1}{3}}{z^3 - z} = \frac{z^2 - \frac{1}{3}}{z(z^2 - 1)} = \frac{z^2 - \frac{1}{3}}{z \ (z-1)(z+1)}$$

$$= \frac{\frac{1}{3}}{z} + \frac{\frac{1}{3}}{z-1} + \frac{\frac{1}{3}}{z+1} \qquad \text{(using partial function)}$$

61

$$\int_c f(z)\partial z = \int_c \frac{\frac{1}{3}}{z}\partial z + \frac{1}{3}\int_c \frac{\partial z}{z-1} + \frac{1}{3}\int_c \frac{\partial z}{z+1}$$

$$= \frac{1}{3}\int_c z^{-1}\partial z + \frac{1}{3}\int_c (z-1)^{-1}\partial z + \frac{1}{3}\int_c (z+1)^{-1}\partial z$$

$$= \frac{1}{3}2\pi j + \frac{1}{3}2\pi j + 0$$

Exercise: Repeat the above Example 2.23 where c:
a)$|z| = 2$  b)$|z - (1 - j)| = \frac{1}{2}$

Example 2.24

Evaluate $\int_c \frac{\partial z}{1+z^3}$  $c: |z + 1| = 1$  counter clock wise.

Solution

Actually, any root of $(1 + z^3)$ i.e inside this circle has integral of $2\pi j$
other wise .It is equal zero.
The root of $(1 + z^3) = 0 \Rightarrow z = \sqrt[3]{-1}$ are

$$\omega_0 = cos\frac{\pi}{3} + j\,sin\frac{\pi}{3} = \frac{1}{2} + j0.866$$

$$\omega_1 = cos\,\pi + j\,sin\,\pi = -1$$

$$\omega_2 = cos\frac{8\pi}{3} + j\,sin\frac{8\pi}{3} = \frac{1}{2} - 0.866$$

We write

$$\frac{1}{z^3 + 1} = \frac{1}{(z + 1)(z^2 - z + 1)}$$

Or $z = -1$ root by $|z + 1| = 0 \rightarrow z = -1$

## 2.11 Evaluation of line integral by indefinite integration

If $f(z)$ is analytic an simply connected domain D, then the exists an indefinite integral of $F(z)$ in the domain D , that is an  analytic function $f(z)$ such that $F'(z) = f(z)$ in D and for all point in D  joining the points a and b in D .

$$\int_a^b f(z)\partial z = F(b) - F(a)$$

Exercise: Repeat the above Example 2.24 where c:

1- $|z| = 2$       2-  $|z - (1 - j)| = \frac{1}{2}$

Example 2.25

Evaluate $\int_c \frac{\partial z}{1+z^3}$   $c: |z + 1| = 1$counterclockwise

Solution:

Any root of $(1 + z^3)$ i.e inside this circle has integral of $2\pi j$,there its equal zero

The root of $(1 + z^3) = 0 \rightarrow z = \sqrt[3]{-1}$ are:

$$\omega_0 = cos\frac{\pi}{3} + j\, sin\frac{\pi}{3} = \frac{1}{2} + j0.866$$

$$\omega_1 = cos\,\pi + j\,sin\,\pi = -1; \omega_2 = cos\frac{3\pi}{3} + j\,sin\frac{3\pi}{3}$$

$$= \frac{1}{2} - j0.866$$

We write :

$$\frac{1}{z^3 + 1} = \frac{1}{(z + 1)(z^2 - z + 1)}$$

$$\Rightarrow \frac{1}{(z + 1)(z - (\frac{1}{2} + j0.866)(z - (\frac{1}{2} - j0.866)}$$

$z = -1$ root

Evaluation of line integral by in define integration

If $F(z)$ is analytic in a simply connected domain D, then there exists an indefinite integral of $f(z)$ in the domain D. That is analytic function $F(z)$ such that $F(z) = f(z)$ in D, and all path in D joining the points $a$ and $b$ in D:

$$\int_a^b F(z)\partial z = F(b) - F(a)$$

Example 2.26 : $\int_j^{1+4j} z^2 \partial z = \left[\frac{z^3}{3}\right]_j^{1+4j} = \frac{1}{3}\{((1+4j)^3 - j^3\} = -\frac{47}{3} - 17j$

Example 2.27: $\left[\int_j^{\frac{\pi}{2}} \cos z\, \partial z = \sin z\right]_j^{\frac{\pi}{2}} = \sin\frac{\pi}{2} - j\sin = 1 - j\sinh j$

Since $\sin z = \sin(\alpha + jy) = \sin\alpha \cdot \cosh y + j\cos\alpha \sinh y$

Exercise: Evaluate $\int_0^{\pi j} e^z \partial z$ Ans $= -z$

## 2.12 Cauchy 's integral formula

Let $f(z)$ be analytic in simply connected domain D, then for any point $z_0$ in D and any simple closed path c in D which enclosed $z_0$ we have

$$\int_c \frac{f(z)}{z - z_0} \partial z = 2\pi j f(z_0)$$

The integration being taken in counter clock wise sense

Example 2.28

Find the value of $\int_c \frac{e^z}{z^2+1} \partial z$ if c is a circle of unit radius with counter of

   a)$z = j$      b)$z = -j$

Solution

a) we can write the integral as

$$\int_c \frac{e^z}{(z\ +j)} \cdot \frac{\partial z}{(z-j)}$$

Comparing with the C.I.F we have $z_0 = j$ which is analytic with and on the above circle [in fact $f(z)$ is analytic every were except at $z_0 = -j$]. Therefore, we can apply C.I.F

$$\int_c \frac{e^z}{(z+j)} \cdot \frac{\partial z}{(z-j)} = 2\pi j f(z_0) = 2\pi j \frac{e^j}{2j} = \pi(\cos 1 + j\sin 1)$$

## 2.13 Residue theorem

Let $f(z)$ be a function which is analytic inside a simple closed path c and on c, except for finitely may singular points $a_1, a_2, \ldots\ldots\ldots a_m$ inside them

$$\int_c F(z)\partial z = 2\pi i \sum_{j=1,z=aj}^{m} Re\, s\, F(z)$$

The integral being taken in the counterclockwise around c where:

$$Re\, s\, F(z) = \lim_{z \to a}\{(z-1)F(z)\}$$

**Theorem 1**

If c is closed curve and if $F(z)$ is analytic within on c except at a finite number of singular points in their of c, them

$$\int_c F(z)\partial z = 2\pi i(r_1 + r_2 + \ldots\ldots r_m)$$

Where $r_1 + r_2 + \ldots\ldots, r_m$ are the residues of $F(z)$ at its singular points within c

Example 2.29: what is the integral of $F(z) = \dfrac{-3z+4}{z(z-1)(z-2)}$ around the circle $c$: $|z| = \dfrac{3}{2}$

Solution:

$$\int_c f(z)\partial z = 2\pi i(r_1 + r_2)$$

$$r_1 = Re\, s\, F(z).\lim_{z=0}\lim_{z \to 0}\left\{ z\,\frac{-3z+4}{z(z-1)(z-2)}\right\} = 2$$

$$r_2 = Re\, s\, F(z).\lim_{z=1}\lim_{z \to 1}\left\{ (z-1)\frac{-3z+4}{z(z-1)(z-2)}\right\} = -1$$

$$\int_e F(z)\partial z = 2\pi i(2+(-j)) = 2\pi i$$

**Theorem 2**

If $F(z)$ has a pole of order $m$ at $z = a$ then residue of $F(z)$ at $z = a$ is

$$r = \frac{1}{(m-1)!}\lim_{z \to a}\left\{\frac{d^{m-1}}{dz^{m-1}}[(z-a)^m F(z)]\right\}$$

Example 2.30

evaluate $\int_c \dfrac{e^z}{(z-1)(z-3)^2}\,\partial z$ where c is

a)   $|z| = \dfrac{3}{2}$

b)   $|z| = 4$

Solution

a)

$$\int_c F(z)\partial z = 2\pi i(r_1)$$

$$r_1 = \operatorname*{Re}_{z=1} s\, F(z) = \lim_{z\to 1}\left\{(z-1).\frac{e^z}{(z-1)(z-3)^2}\right\} = \frac{e}{4}$$

$$\int_c F(z)\partial z = 2\pi i\,\frac{e}{4} = \frac{e\pi i}{2}$$

b) $|z| = 4$

$$\int_c F(z)\partial z = 2\pi i(r_1 + r_2)$$

$r_1 = \frac{e}{4}$ from part (a)

$$r_2 = \frac{1}{z-2}\lim_{z\to 3}\left\{\frac{\partial}{\partial z}\left\{(z-3)^2.\frac{e^z}{(z-1)(z-3)^2}\right\}\right\}$$

$$= \lim_{z\to 3}\left\{\frac{\partial}{\partial z}\left(\frac{e^z}{z-1}\right)\right\}$$

$$= \lim_{z\to 3}\left[\frac{(z-1)e^z.e^z}{(z-1)^2}\right] = \frac{e^3}{4}$$

$$r_2 = \frac{e^3}{4}$$

$$\int_c F(z)\partial z = 2\pi i\left(\frac{e}{4} + \frac{e^3}{4}\right)$$

Example 2.31

Integrate $F(z) = \frac{1}{(z^3-1)^2}$ in the C.Csense around $c: |z-1| = 1$ (at $z = 1$center of circle)

Solution

$$z^3 - 1 = 0 \Rightarrow z^3 = 1 \Rightarrow z = \sqrt[3]{1}$$

$$zk = r^{\frac{1}{n}}e^{j(\frac{\theta+2\pi k}{n})}$$

$$z_0 = 1$$

$$z_1 = e^{\frac{2\pi}{3}j} = \cos\frac{2\pi}{3} + j\sin\frac{2\pi}{3} = -0.5 + j0.866$$

$$z_2 = e^{\frac{-2\pi}{3}j} = \cos\frac{2\pi}{3} - j\sin\frac{2\pi}{3} = -0.5 - j0.866$$

$$(z^3 - 1) = (z - z_0)(z - z_1)(z - z_2)$$

$$= (z - 1)(z + 0.5 - j0.866)(z + 0.5 + j0.866)$$

$$F(z) = \frac{1}{(z^3 - 1)^2} = \frac{1}{[(z - 1)(z^2 + z + 0)]^2}$$

$$F(z) = \frac{1}{(z - 1)(z^2 + z + 1)}$$

$$\int F(z)dz = 2\pi i(r_i)$$

$$r_1 = \frac{1}{(z - 1)!}\lim_{z \to 1}\left\{\frac{d}{dz}\left[(z - 1)^2 \cdot \frac{1}{(z - 1)^2(z^2 + z + 1)^2}\right]\right\}$$

$$= \lim_{z \to 1}\left\{\frac{d}{dz}\left[\frac{1}{(z^2 + z + 1)^2}\right]\right\} = \lim_{z \to 1}\left\{\frac{-2(z^2 + z + 1)(2z + 1)}{(z^2 + z + 1)^2}\right\}$$

$$\int_c F(z)\,\partial z = 2\pi i(\frac{-2}{3}) = \frac{-4\pi i}{3}$$

Example 2.32

Using the theory of residue integrant $F(z) = \frac{ze^z}{(z-1)^3}$ around a simple counter $c: |z| = 1$

Solution

$$r_1 = \frac{1}{2!}\lim_{z \to 1}\left\{\frac{d^2}{dz^2}(z - 1)^3\frac{ze^z}{(z - 1)^3}\right\}$$

$$= \frac{1}{2}\lim_{z \to 1}\frac{d^2}{dz^2}(ze^z)$$

$$= \frac{1}{2}\lim_{z \to 1}\frac{d}{dz}[ze^z + e^z] = \frac{1}{2}\lim_{z \to 1}[ze^z + ze^z] = \frac{3e}{2}$$

$$\int_c F(z)dz = 2\pi j \frac{3e}{2} = 3\pi je$$

**Problems**

2.1 If $Z_1 = 3 + 3\,j, Z_2 = -15 + 10\,j$
Find $Z_1 + Z_2$ *and* $Z_1 - Z_2$

2.2 If $Z = 2 + j4$ **Find** $r, \theta$ and $Z$

2.3 Find the value of $e^{10x}$ using Euler's Formula

2.4 Find the value of $e^{-2x}$ using Euler's Formula

2.5 Find the value of $e^{-0.5x}$ using Euler's Formula

2.6 Solve the equation $Z^4 = -81$

2.7 Solve the equation $Z^4 = -100$

2.8 Describe the set of points
$|Z - 3| = 2$
$1 \le |Z - 2| \le 4$

2.9 Show that if
$f(z) = e^{2z}$ is analytic or not, and find $f'(z)$

2.10 Show that if
$f(z) = e^{-z} + 1$ is analytic or not, and find $f'(z)$

2.11 Find the conjugate the harmonic function of the harmonic function $u = 2\alpha^2 - 4y^2$

# Chapter 3

## Matrix Algebra

Matrix algebra leads to some applications, such as geometrical transformations (essential to image processing, one of many engineering applications) and evolutionary models (economics, probability - Markov chains, etc.).

### 3.1 Simple Matrix Algebra

A matrix with $m$ rows and $n$ columns has **dimensions** or **size** ($m \times n$) and is said to be an "$m$ by $n$ matrix". The number of rows is always written first and the number of columns second.

An example of a 2×3 matrix is $A = \begin{bmatrix} 3 & 2 & 1 \\ 0 & -1 & -2 \end{bmatrix}$.

A 1×$n$ matrix is a **row matrix**. $R = \begin{bmatrix} 12 & 2 & -5 & 10 \end{bmatrix}$ is a row matrix (of size 1×4).
(also known as a **row vector**).

An $n \times 1$ matrix is a **column matrix**. $C = \begin{bmatrix} -2 \\ 0 \\ 1 \end{bmatrix}$ is a column matrix (of size 3×1).
(also known as a **column vector**).

A matrix with equal numbers of rows and columns is a **square matrix**.

$D = \begin{bmatrix} -1 & 0 & 2 \\ 0 & 1 & 3 \\ 0 & 0 & 5 \end{bmatrix}$ is a square matrix of dimensions (3×3).

The entry in row $i$ and column $j$ of matrix $D$ is $d_{ij}$.
In matrix $D$ above, $d_{23} = 3$.

The **main diagonal** of a matrix extends down and right from the top left corner; the elements of the main diagonal of matrix $A = [a_{ij}]$ are $a_{ij}$.

For the four matrices above, the main diagonals are highlighted here:

$$A = \begin{bmatrix} \boxed{3} & 2 & 1 \\ 0 & \boxed{-1} & -2 \end{bmatrix}, \qquad R = \begin{bmatrix} \boxed{12} & 2 & -5 & 10 \end{bmatrix},$$

$$C = \begin{bmatrix} \boxed{-2} \\ 0 \\ 1 \end{bmatrix},$$

$$D = \begin{bmatrix} \boxed{-1} & 0 & 2 \\ 0 & \boxed{1} & 3 \\ 0 & 0 & \boxed{5} \end{bmatrix}.$$

## Equality

Two matrices are equal if and only if they are the same size *and* all corresponding pairs of entries are equal.

In other words, $A = B$ iff $a_{ij} = b_{ij}$ for all $i$ and for all $j$.

Example:

$$\begin{bmatrix} a & b \\ c & d \end{bmatrix} = \begin{bmatrix} 2 & 0 \\ 1 & 3 \end{bmatrix} \quad \Leftrightarrow \quad a = 2, \quad b = 0, \quad c = 1 \text{ and } d = 3$$

**Addition** is defined only for matrices of the same size.

Example 3.1

$$A = \begin{bmatrix} 0 & 1 \\ 3 & 2 \\ -1 & 0 \end{bmatrix}, \quad B = \begin{bmatrix} 6 & 5 \\ 4 & 3 \\ 2 & 1 \end{bmatrix} \quad \Rightarrow \quad A + B = \begin{bmatrix} 6 & 6 \\ 7 & 5 \\ 1 & 1 \end{bmatrix}$$

Example 3.2

$$A = \begin{bmatrix} 0 & 1 \\ 3 & 2 \\ -1 & 0 \end{bmatrix}, \quad C = \begin{bmatrix} 6 & 5 \\ 2 & 1 \end{bmatrix} \quad \Rightarrow \quad A + C \text{ is unde-}$$

fined.

Matrix addition is commutative and associative. For any matrices $A$, $B$, and $C$ of the same size,

$$A + B = B + A \text{ and}$$
$$A + (B + C) = (A + B) + C$$

70

The identity matrix under addition is the **zero matrix**:

All entries of any zero matrix are zero.    The $(m{\times}n)$ zero matrix is $O_{mn}$ (or just $O$ if the size is obvious from the situation).

For all matrices $X$,
$$X + O = X \qquad \text{(where the zero matrix is the same size as } X)$$

The inverse matrix of an $(m{\times}n)$ matrix $A$ under addition is its **negative** $-A$, whose entries are all $-a_{ij}$.

For all matrices $X$,
$$X + (-X) = O \quad \text{(where the zero matrix is the same size as } X)$$

The **difference** of two matrices $A$, $B$ of the same size is
$$A - B = A + (-B), \text{ whose elements are } [a_{ij} - b_{ij}]$$

Example 3.3

$$A = \begin{bmatrix} 0 & 1 \\ 3 & 2 \\ -1 & 0 \end{bmatrix}, \quad B = \begin{bmatrix} 6 & 5 \\ 4 & 3 \\ 2 & 1 \end{bmatrix} \quad \Rightarrow \quad B - A = \begin{bmatrix} 6 & 4 \\ 1 & 1 \\ 3 & 1 \end{bmatrix}$$

Example 3.4

$$A = \begin{bmatrix} 0 & 1 \\ 3 & 2 \\ -1 & 0 \end{bmatrix}, \quad C = \begin{bmatrix} 6 & 5 \\ 2 & 1 \end{bmatrix} \quad \Rightarrow \quad A - C \text{ is unde-}$$
fined.

## 3.2 Scalar Multiplication

Multiplication of a matrix $A$ by a scalar $k$ causes every element of $A$ to be multiplied by $k$.
$$A = [a_{ij}] \quad \Rightarrow \quad kA = [ka_{ij}]$$

Example 3.5

$$A = \begin{bmatrix} 0 & 1 \\ 3 & 2 \\ -1 & 0 \end{bmatrix}, \quad B = \begin{bmatrix} 6 & 5 \\ 4 & 3 \\ 2 & 1 \end{bmatrix} \quad \Rightarrow \quad 2A = \begin{bmatrix} 0 & 2 \\ 6 & 4 \\ -2 & 0 \end{bmatrix}$$

$$\text{and} \quad 3A - 2B = \begin{bmatrix} 3(0)-2(6) & 3(1)-2(5) \\ 3(3)-2(4) & 3(2)-2(3) \\ 3(-1)-2(2) & 3(0)-2(1) \end{bmatrix} = \begin{bmatrix} -12 & -7 \\ 1 & 0 \\ -7 & -2 \end{bmatrix}$$

The **distributive** laws for matrices of the same size follow:

$$k(A+B) = kA + kB$$
$$(p+q)A = pA + qA$$
and
$$(pq)A = p(qA)$$

The **transpose** of a matrix $A = [\, a_{ij}\,]$ is $A^T = [\, a_{ji}\,]$.

Thus the rows of the transpose are the columns of the original matrix and vice versa.
The transpose of an $(m \times n)$ matrix is an $(n \times m)$ matrix.
In particular, the transpose of a row matrix is a column matrix and the transpose of a column matrix is a row matrix.

Example 3.6

Write down the transpose of the following matrices:

$$A = \begin{bmatrix} 1 & 2 \\ 3 & 4 \end{bmatrix}, \quad B = \begin{bmatrix} 9 \\ -6 \\ 1 \\ 2 \end{bmatrix}, \quad C = \begin{bmatrix} 1 & 3 & 6 \\ -4 & 0 & 5 \end{bmatrix}$$

$$A^T = \begin{bmatrix} 1 & 3 \\ 2 & 4 \end{bmatrix}, \quad B^T = [\, 9 \quad -6 \quad 1 \quad 2\,], \quad C^T = \begin{bmatrix} 1 & -4 \\ 3 & 0 \\ 6 & 5 \end{bmatrix}$$

Further properties of transposition:
For all equal-size matrices $A, B$ and all scalars $k$,

$$(A^T)^T = A$$
$$(kA)^T = k(A^T)$$
$$(A + B)^T = A^T + B^T$$

A matrix for which $A^T = A$ is **symmetric**.
Symmetric matrices are necessarily square $(n \times n)$
and the main diagonal is a line of symmetry..

<u>Example 3.7</u>

$$A = \begin{bmatrix} 1 & -2 \\ -2 & 4 \end{bmatrix}, \quad B = \begin{bmatrix} 1 & -2 \\ -3 & 4 \end{bmatrix}, \quad C = \begin{bmatrix} 1 & -4 & 0 \\ -4 & 2 & 0 \end{bmatrix}$$

Matrix $A$ is symmetric because $A^T = A$ .
Matrix $B$ is not symmetric because
$$B^T = \begin{bmatrix} 1 & -3 \\ -2 & 4 \end{bmatrix} \neq \begin{bmatrix} 1 & -2 \\ -3 & 4 \end{bmatrix} = B.$$
Matrix $C$ cannot be symmetric because it is not square.

## Miscellaneous Examples

<u>Example 3.8</u>   Textbook exercises 3 page 34 question 1(b)

Find $a, b, c$ and $d$ if $\begin{bmatrix} a - b & b - c \\ c - d & d - a \end{bmatrix} = 2 \begin{bmatrix} 1 & 1 \\ -3 & 1 \end{bmatrix}.$

This generates the system of simultaneous linear equations
$$a - b = 2$$
$$b - c = 2$$
$$c - d = -6$$
$$-a \quad + d = 2$$

Solving the linear system,

$$\begin{bmatrix} 1 & -1 & 0 & 0 & 2 \\ 0 & 1 & -1 & 0 & 2 \\ 0 & 0 & 1 & -1 & -6 \\ -1 & 0 & 0 & 1 & 2 \end{bmatrix} \xrightarrow{R_4 + R_1} \begin{bmatrix} 1 & -1 & 0 & 0 & 2 \\ 0 & 1 & -1 & 0 & 2 \\ 0 & 0 & 1 & -1 & -6 \\ 0 & -1 & 0 & 1 & 4 \end{bmatrix}$$

$$\xrightarrow{R_4 + R_2} \begin{bmatrix} 1 & -1 & 0 & 0 & 2 \\ 0 & 1 & -1 & 0 & 2 \\ 0 & 0 & 1 & -1 & -6 \\ 0 & 0 & -1 & 1 & 6 \end{bmatrix} \xrightarrow{R_4 + R_3} \begin{bmatrix} 1 & -1 & 0 & 0 & 2 \\ 0 & 1 & -1 & 0 & 2 \\ 0 & 0 & 1 & -1 & -6 \\ 0 & 0 & 0 & 0 & 0 \end{bmatrix}$$

which is row-echelon form.

$d$ is a non-leading variable and is assigned a parametric value $t$ (where $t$ may be any real number).

The system is now

$$a - b \qquad\qquad = 2$$
$$\qquad b - c \qquad = 2$$
$$\qquad\qquad c - d = -6$$
$$\qquad\qquad\qquad d = t$$

Using back-substitution,

$$c = t - 6$$
$$b = c + 2 = t - 4$$
$$a = b + 2 = t - 2$$

The values of $a, b, c$ and $d$ are therefore
$(a, b, c, d) = (t - 2, t - 4, t - 6, t)$ or equivalently
$(a, b, c, d) = (-2, -4, -6, 0) + t\,(1, 1, 1, 1)$, $(t \in \mathbb{R})$.

Example 3.9

Find the transpose of $A = \begin{bmatrix} 0 & 5 & -2 \\ -5 & 0 & -1 \\ 2 & 1 & 0 \end{bmatrix}$.

$$A^T = \begin{bmatrix} 0 & -5 & 2 \\ 5 & 0 & 1 \\ -2 & -1 & 0 \end{bmatrix} = -A$$

Matrices which are such that $A^T = -A$ are **skew-symmetric**.
In any skew-symmetric matrix $A$, the main diagonal elements $a_{ii} = 0$.

Example 3.10   Textbook exercises 3 page 35 question 15(a)

Find the matrix $A$ that satisfies the equation

$$\left( A + 3\begin{bmatrix} 1 & -1 & 0 \\ 1 & 2 & 4 \end{bmatrix} \right)^T = \begin{bmatrix} 2 & 1 \\ 0 & 5 \\ 3 & 8 \end{bmatrix}$$

Method 1.

$$A + 3\begin{bmatrix} 1 & -1 & 0 \\ 1 & 2 & 4 \end{bmatrix} = \begin{bmatrix} 2 & 1 \\ 0 & 5 \\ 3 & 8 \end{bmatrix}^T = \begin{bmatrix} 2 & 0 & 3 \\ 1 & 5 & 8 \end{bmatrix}$$

$$\Rightarrow A = \begin{bmatrix} 2 & 0 & 3 \\ 1 & 5 & 8 \end{bmatrix} - 3\begin{bmatrix} 1 & -1 & 0 \\ 1 & 2 & 4 \end{bmatrix} = \underline{\begin{bmatrix} -1 & 3 & 3 \\ -2 & -1 & -4 \end{bmatrix}}$$

Method 2.

$$A^T + 3\begin{bmatrix} 1 & 1 \\ -1 & 2 \\ 0 & 4 \end{bmatrix} = \begin{bmatrix} 2 & 1 \\ 0 & 5 \\ 3 & 8 \end{bmatrix}$$

$$\Rightarrow A^T = \begin{bmatrix} 2 & 1 \\ 0 & 5 \\ 3 & 8 \end{bmatrix} - 3\begin{bmatrix} 1 & 1 \\ -1 & 2 \\ 0 & 4 \end{bmatrix} = \begin{bmatrix} -1 & -2 \\ 3 & -1 \\ 3 & -4 \end{bmatrix}$$

$$\Rightarrow A = \begin{bmatrix} -1 & -2 \\ 3 & -1 \\ 3 & -4 \end{bmatrix}^T = \underline{\begin{bmatrix} -1 & 3 & 3 \\ -2 & -1 & -4 \end{bmatrix}}$$

<u>Example 3.11</u>  Textbook exercises 3 page 35 question 17

Show that $A + A^T$ is symmetric for *any* square matrix $A$.

First note that if $A$ is not square, then the dimensions of $A$ and $A^T$ will be different, so that $A + A^T$ is not defined at all.

$(A + A^T)^T = A^T + A = A + A^T$  (matrix addition is commutative).

Therefore the matrix  $(A + A^T)$  is symmetric for all square matrices $A$.

Building on this example, any square matrix $A$ can be written as the sum of a symmetric matrix $S$ and a skew-symmetric matrix $K$:  $A = S + K$

$S$ is symmetric  $\Rightarrow S^T = S$.

75

$K$ is skew-symmetric $\Rightarrow K^T = -K$.

$A^T = (S + K)^T = S^T + K^T = S - K$

$$\Rightarrow \quad A + A^T = (S + K) + (S - K) = 2S \quad \Rightarrow \quad S = \frac{A + A^T}{2}$$

and

$$A - A^T = (S + K) - (S - K) = 2K \quad \Rightarrow \quad K = \frac{A - A^T}{2}$$

so that the symmetric matrix $S$ and the skew-symmetric matrix $K$ are uniquely determined for each square matrix $A$. [This is also question 20, exercise 3, on page 36 of the textbook.]

## 3.3 Matrix Multiplication

### Dot product

The dot product of a row vector $R = \begin{bmatrix} r_1 & r_2 & \cdots & r_n \end{bmatrix}$

and a column vector $C = \begin{bmatrix} c_1 \\ c_2 \\ \vdots \\ c_n \end{bmatrix} = \begin{bmatrix} c_1 & c_2 & \cdots & c_n \end{bmatrix}^T$

is defined to be

$$R \odot C = \begin{bmatrix} r_1 & r_2 & \cdots & r_n \end{bmatrix} \begin{bmatrix} c_1 \\ c_2 \\ \vdots \\ c_n \end{bmatrix} = r_1 c_1 + r_2 c_2 + \cdots + r_n c_n =$$

$\sum_{k=1}^{n} r_k c_k$

Note that the dimensions of the row and column vectors must be $(1 \times n)$ and $(n \times 1)$ respectively, otherwise the sum $R \odot C = \sum_{k=1}^{n} r_k c_k$ is not defined.

The order of multiplication in the dot product is important.

### Example 3.12

The numbers of atoms of carbon, hydrogen and oxygen in each molecule of water, methanol and ethanol are represented in the matrix $A$:

$$A = \begin{array}{c} \\ C \\ H \\ O \end{array} \begin{array}{ccc} \text{water} & \text{methanol} & \text{ethanol} \\ \left[\begin{array}{ccc} 0 & 1 & 2 \\ 2 & 4 & 6 \\ 1 & 1 & 1 \end{array}\right] \end{array}$$

The composition of a form of denatured alcohol and a dilution of that alcohol in water is described by the numbers of molecules of water, methanol and ethanol per 20 molecules of the alcohol, as listed in matrix $B$:

denatured diluted

$$B = \begin{array}{c} \text{water} \\ \text{methanol} \\ \text{ethanol} \end{array} \left[\begin{array}{cc} 0 & 10 \\ 2 & 1 \\ 18 & 9 \end{array}\right]$$

Find the ratio of carbon atoms to hydrogen atoms to oxygen atoms in the diluted alcohol.

Every 20 molecules on average of the diluted alcohol contains 10 molecules of water, 1 molecule of methanol and 9 molecules of ethanol. Reading the atomic contents of these three molecules from matrix $A$, we find the numbers of atoms per 20 molecules of diluted alcohol to be:

|  | Water + Meth. + Ethanol |
|---|---|
| Carbon: | $0 \times 10 + 1 \times 1 + 2 \times 9 = 19$ atoms |
| Hydrogen: | $2 \times 10 + 4 \times 1 + 6 \times 9 = 78$ atoms |
| Oxygen: | $1 \times 10 + 1 \times 1 + 1 \times 9 = 20$ atoms |

On average, every 20 molecules of diluted alcohol contain 19 atoms of carbon, 78 atoms of hydrogen and 20 atoms of oxygen. The ratio is C:H:O: = **19 : 78 : 20**.

Note how the numbers of atoms were found.
The number of carbon atoms is the dot product of the first row of $A$ with the second column of $B$.
The number of hydrogen atoms is the dot product of the second row of $A$ with the second column of $B$.
The number of oxygen atoms is the dot product of the third row of $A$ with the second column of $B$.

The product of the two matrices yields the number of atoms per 20 molecules of each of the two substances:

$$AB = \begin{bmatrix} 0 & 1 & 2 \\ 2 & 4 & 6 \\ 1 & 1 & 1 \end{bmatrix} \boxed{} \begin{bmatrix} 0 & 10 \\ 2 & 1 \\ 18 & 9 \end{bmatrix}$$

$$= \begin{bmatrix} (0 \times 0 + 1 \times 2 + 2 \times 18) & (0 \times 10 + 1 \times 1 + 2 \times 9) \\ (2 \times 0 + 4 \times 2 + 6 \times 18) & (2 \times 10 + 4 \times 1 + 6 \times 9) \\ (1 \times 0 + 1 \times 2 + 1 \times 18) & (1 \times 10 + 1 \times 1 + 1 \times 9) \end{bmatrix}$$

denatured diluted

$$\Rightarrow \quad AB = \begin{bmatrix} 38 & 19 \\ 116 & 78 \\ 20 & 20 \end{bmatrix} \begin{matrix} C \\ H \\ O \end{matrix}$$

The product of two general matrices follows.

The product of an $(m \times n)$ matrix $A$ with a $(p \times q)$ matrix $B$ (in that order) is not defined unless $p = n$.

The product $C = AB$ of an $(m \times n)$ matrix $A$ with an $(n \times q)$ matrix $B$ (in that order) is the $(m \times q)$ matrix $C = [\, c_{ij} \,]$, where the entry in row $i$ and column $j$ of $C$ is the dot product of the $i$th row of $A$ with the $j$th column of $B$:

$$c_{ij} = \sum_{k=1}^{n} a_{ik} c_{kj}.$$

Example 3.13

Find the matrix products $AB$ and $BA$ where

$$A = \begin{bmatrix} 1 & 0 \\ -2 & 1 \\ 4 & 3 \end{bmatrix} \quad \text{and} \quad B = \begin{bmatrix} 2 & 1 \\ 1 & -2 \end{bmatrix}.$$

$$AB = \begin{bmatrix} 1 & 0 \\ -2 & 1 \\ 4 & 3 \end{bmatrix} \begin{bmatrix} 2 & 1 \\ 1 & -2 \end{bmatrix} =$$

$$\begin{bmatrix} (1 \times 2 + 0 \times 1) & (1 \times 1 + 0 \times -2) \\ (-2 \times 2 + 1 \times 1) & (-2 \times 1 + 1 \times -2) \\ (4 \times 2 + 3 \times 1) & (4 \times 1 + 3 \times -2) \end{bmatrix} = \begin{bmatrix} 2 & 1 \\ -3 & -4 \\ 11 & -2 \end{bmatrix}$$

$BA$ is **not defined** because $B$ is (2×2) and $A$ is (3×2). The number of columns of the left matrix does not match the number of rows of the right matrix.

Note that this example demonstrates that matrix multiplication is **not commutative** in general, that is $BA \neq AB$.

The **identity matrix** of order $n$ is the square ($n \times n$) matrix whose main diagonal entries are one and whose other entries are all zero.

$$I_2 = \begin{bmatrix} 1 & 0 \\ 0 & 1 \end{bmatrix} \quad \text{and} \quad I_3 = \begin{bmatrix} 1 & 0 & 0 \\ 0 & 1 & 0 \\ 0 & 0 & 1 \end{bmatrix}, \text{etc.}$$

For any ($m \times n$) matrix $A$, $I_m A = A I_n = A$.
$I$ is therefore the identity element for the operation of matrix multiplication.

Where it is obvious from the context, $I_n$ is represented by just $I$.

Example 3.14

$$A = \begin{bmatrix} 1 & 0 \\ -2 & 1 \\ 4 & 3 \end{bmatrix} \Rightarrow IA = \begin{bmatrix} 1 & 0 & 0 \\ 0 & 1 & 0 \\ 0 & 0 & 1 \end{bmatrix}\begin{bmatrix} 1 & 0 \\ -2 & 1 \\ 4 & 3 \end{bmatrix} = \begin{bmatrix} 1 & 0 \\ -2 & 1 \\ 4 & 3 \end{bmatrix}$$

and

$$AI = \begin{bmatrix} 1 & 0 \\ -2 & 1 \\ 4 & 3 \end{bmatrix}\begin{bmatrix} 1 & 0 \\ 0 & 1 \end{bmatrix} = \begin{bmatrix} 1 & 0 \\ -2 & 1 \\ 4 & 3 \end{bmatrix}$$

Where the product is defined, the product of the **zero matrix** with any other matrix is the zero matrix of the appropriate dimensions.

Example 3.15

$$\begin{bmatrix} 0 & 0 & 0 \\ 0 & 0 & 0 \end{bmatrix}\begin{bmatrix} 1 & 0 \\ -2 & 1 \\ 4 & 3 \end{bmatrix} = \begin{bmatrix} 0 & 0 \\ 0 & 0 \end{bmatrix}, \qquad \begin{bmatrix} 1 & 0 \\ -2 & 1 \\ 4 & 3 \end{bmatrix}\begin{bmatrix} 0 & 0 & 0 & 0 \\ 0 & 0 & 0 & 0 \end{bmatrix}$$
$$= \begin{bmatrix} 0 & 0 & 0 & 0 \\ 0 & 0 & 0 & 0 \\ 0 & 0 & 0 & 0 \end{bmatrix}$$

but $\begin{bmatrix} 0 & 0 \\ 0 & 0 \end{bmatrix} \begin{bmatrix} 1 & 0 \\ -2 & 1 \\ 4 & 3 \end{bmatrix}$ is not defined

Example 3.16

Find $A^2$, where $A = \begin{bmatrix} 1 & 2 \\ 3 & 4 \end{bmatrix}$.

$$A^2 = \begin{bmatrix} 1 & 2 \\ 3 & 4 \end{bmatrix} \begin{bmatrix} 1 & 2 \\ 3 & 4 \end{bmatrix} = \begin{bmatrix} (1 \times 1 + 2 \times 3) & (1 \times 2 + 2 \times 4) \\ (3 \times 1 + 4 \times 3) & (3 \times 2 + 4 \times 4) \end{bmatrix}$$
$$= \begin{bmatrix} 7 & 10 \\ 15 & 22 \end{bmatrix}$$

Example 3.17

Find $A^2$, where $A = \begin{bmatrix} -1 & k \\ 0 & 1 \end{bmatrix}$ and $k$ is any real number.

$$A^2 = \begin{bmatrix} -1 & k \\ 0 & 1 \end{bmatrix} \begin{bmatrix} -1 & k \\ 0 & 1 \end{bmatrix} = \begin{bmatrix} (-1 \times -1 + k \times 0) & (-1 \times k + k \times 1) \\ (0 \times -1 + 1 \times 0) & (0 \times k + 1 \times 1) \end{bmatrix}$$
$$= \begin{bmatrix} 1 & 0 \\ 0 & 1 \end{bmatrix} = I$$

Note that in scalar arithmetic $x^2 = 1 \Rightarrow x = \pm 1$, but in matrix multiplication
$A^2 = I \quad \boxed{\Rightarrow} \quad A = \pm I$

Example 3.18

Find $A^2$, where $A = \begin{bmatrix} 2 & -4 \\ 1 & -2 \end{bmatrix}$.

$$A^2 = \begin{bmatrix} 2 & -4 \\ 1 & -2 \end{bmatrix} \begin{bmatrix} 2 & -4 \\ 1 & -2 \end{bmatrix}$$
$$= \begin{bmatrix} (2 \times 2 + -4 \times 1) & (2 \times -4 + -4 \times -2) \\ (1 \times 2 + -2 \times 1) & (1 \times -4 + -2 \times -2) \end{bmatrix} = \begin{bmatrix} 0 & 0 \\ 0 & 0 \end{bmatrix}$$
$$= 0$$

Note that in scalar arithmetic $x^2 = 0 \Rightarrow x = 0$, but in matrix multiplication
$A^2 = 0 \quad \boxed{\Rightarrow} \quad A = 0$

## 3.4 properties of matrix multiplication

For any scalar $k$, matrices $A$, $B$, $C$ of dimensions such that the matrix multiplications are defined, and identity and zero matrices of the appropriate dimensions,

$IA = AI = A$          [identity]
$OA = AO = O$   [zero]
$A(BC) = (AB)C$      [associative law]
$A(B+C) = AB + AC$    [distributive law]
$(B+C)A = BA + CA$    [distributive law]
$k(AB) = (kA)B = A(kB)$
but note that $AB \neq BA$ in general. Matrices for which $AB = BA$ are said to commute.
Be very careful of the order of matrix multiplication.

$(AB)^\mathsf{T} = B^\mathsf{T}A^\mathsf{T}$

As first seen in Chapter 1, any system of linear equations

$$a_{11}x_1 + a_{12}x_2 + a_{13}x_3 + \ldots + a_{1n}x_n = b_1$$
$$a_{21}x_1 + a_{22}x_2 + a_{23}x_3 + \ldots + a_{2n}x_n = b_2$$
$$a_{31}x_1 + a_{32}x_2 + a_{33}x_3 + \ldots + a_{3n}x_n = b_3$$
$$\vdots$$
$$a_{p1}x_1 + a_{p2}x_2 + a_{p3}x_3 + \ldots + a_{pn}x_n = b_p$$

can be written more compactly as the matrix equation

$$AX = B$$

$$\text{where } A = \begin{bmatrix} a_{ij} \end{bmatrix} = \begin{bmatrix} a_{11} & a_{12} & a_{13} & \cdots & a_{1n} \\ a_{21} & a_{22} & a_{23} & \cdots & a_{2n} \\ a_{31} & a_{32} & a_{33} & \cdots & a_{3n} \\ \vdots & \vdots & \vdots & \ddots & \vdots \\ a_{p1} & a_{p2} & a_{p3} & \cdots & a_{pn} \end{bmatrix},$$

and $X$ and $B$ are the column vectors

$$X = \begin{bmatrix} x_1 \\ x_2 \\ x_3 \\ \vdots \\ x_n \end{bmatrix}, \quad B = \begin{bmatrix} b_1 \\ b_2 \\ b_3 \\ \vdots \\ b_p \end{bmatrix}.$$

Given an inhomogeneous linear system $AX = B$, there is an **associated homogeneous system**

$$AX = 0$$

If the column vector $X_1$ is any one solution to $AX = B$ and the column vector $X_0$ is any one solution to $AX = 0$, then $(X_0 + X_1)$ is also a solution to $AX = B$. [This requires $AX = B$ to be consistent.]

Thus the general solution to the system $AX = B$ may be expressed as the sum of the general solution to the associated homogeneous system and a **particular solution** of the inhomogeneous system.

Proof:

Let $X_2$ be any solution to $AX = B$ (so that $AX_2 = B$) and let $X_1$ be a known particular solution to $AX = B$ (so that $AX_1 = B$).
Let $X_0 = X_2 - X_1$.

Then
$$AX_0 = A(X_2 - X_1) = AX_2 - AX_1 = B - B = 0$$
$\Rightarrow$ $X_0$ is a solution to the associated homogeneous system $AX = 0$.

Occasionally it is easier to find a particular solution and to solve the associated homogeneous system than it is to solve the original inhomogeneous system all at once.

We will see this concept of partitioning a solution into a particular solution and the solution of the associated homogeneous system again when we study ordinary differential equations in a future course (MATH 3260 or ENGI 3424 or ENGI 3425/4425).

If $A$ is an $(m \times n)$ matrix of rank $r$, then the homogeneous linear system of $m$ equation in $n$ variables $AX = 0$ has exactly $(n-r)$ basic solutions, one for each parameter and every solution is a linear combination of these basic solutions.

## Example 3.19

Find basic solutions of $AX = 0$, where

$$A = \begin{bmatrix} 1 & 2 & 0 & 1 & 1 \\ 2 & 4 & 1 & 1 & -2 \\ 3 & 6 & 1 & 2 & -1 \\ 1 & 2 & 1 & 0 & -3 \end{bmatrix}$$

Show that $X = \begin{bmatrix} 1 & 2 & 0 & 2 & 1 \end{bmatrix}^T$ is a solution to $AX = B$, where

$B = \begin{bmatrix} 8 & 10 & 18 & 2 \end{bmatrix}^T$. Hence find the complete solution to $AX = B$.

Reducing the augmented matrix to row-echelon form:

$$\begin{bmatrix} 1 & 2 & 0 & 1 & 1 & 0 \\ 2 & 4 & 1 & 1 & -2 & 0 \\ 3 & 6 & 1 & 2 & -1 & 0 \\ 1 & 2 & 1 & 0 & -3 & 0 \end{bmatrix} \xrightarrow[R_4-R_1]{\substack{R_3-3R_1 \\ R_2-2R_1}} \begin{bmatrix} 1 & 2 & 0 & 1 & 1 & 0 \\ 0 & 0 & 1 & -1 & -4 & 0 \\ 0 & 0 & 1 & -1 & -4 & 0 \\ 0 & 0 & 1 & -1 & -4 & 0 \end{bmatrix}$$

$$\xrightarrow{R_3-R_2 \, R_4-R_2} \begin{bmatrix} \boxed{1} & 2 & 0 & 1 & 1 & 0 \\ 0 & 0 & \boxed{1} & -1 & -4 & 0 \\ 0 & 0 & 0 & 0 & 0 & 0 \\ 0 & 0 & 0 & 0 & 0 & 0 \end{bmatrix}$$

which is equivalent to
$$x_1 + 2x_2 + 0x_3 + x_4 + x_5 = 0 \quad \text{and}$$
$$x_3 - x_4 - 4x_5 = 0$$
The leading variables are $x_1$ and $x_3$.
Assign parameters $x_2 = r$, $x_4 = s$, $x_5 = t$, so that the general solution is
$$x_1 = -2r - s - t, \quad x_3 = s + 4t$$
Then

$$X = \begin{bmatrix} x_1 \\ x_2 \\ x_3 \\ x_4 \\ x_5 \end{bmatrix} = \begin{bmatrix} -2r - s - t \\ r \\ s + 4t \\ s \\ t \end{bmatrix} = \begin{bmatrix} -2 \\ 1 \\ 0 \\ 0 \\ 0 \end{bmatrix} r + \begin{bmatrix} -1 \\ 0 \\ 1 \\ 1 \\ 0 \end{bmatrix} s + \begin{bmatrix} -1 \\ 0 \\ 4 \\ 0 \\ 1 \end{bmatrix} t$$

The basic solutions are therefore

83

$X_1 = [-2\ 1\ 0\ 0\ 0]^T,\ X_2 = [-1\ 0\ 1\ 1\ 0]^T,\ X_3 = [-1\ 0\ 4\ 0\ 1]^T$

and the general solution to $AX = O$ is $X = r X_1 + s X_2 + t X_3$.

$$
AX = \begin{bmatrix} 1 & 2 & 0 & 1 & 1 \\ 2 & 4 & 1 & 1 & -2 \\ 3 & 6 & 1 & 2 & -1 \\ 1 & 2 & 1 & 0 & -3 \end{bmatrix} \begin{bmatrix} 1 \\ 2 \\ 0 \\ 2 \\ 1 \end{bmatrix} = \begin{bmatrix} 8 \\ 10 \\ 18 \\ 2 \end{bmatrix} = B
$$

Therefore the complete solution to $AX = B$ is
$$X = [1\ 2\ 0\ 2\ 1]^T + r X_1 + s X_2 + t X_3.$$

## 3.5 Block multiplication

<u>Example 3.20</u>

Suppose that matrices $A, B, P, X$ and $Y$ are defined as

$$
A = \begin{bmatrix} 1 & 0 & 0 & 0 & 0 \\ 0 & 1 & 0 & 0 & 0 \\ 0 & 0 & 2 & 0 & 3 \\ 0 & 0 & 0 & 3 & 1 \end{bmatrix} = \begin{bmatrix} I_2 & O_{23} \\ O_{22} & P \end{bmatrix} \text{ and } B = \begin{bmatrix} 1 & 2 \\ 3 & 4 \\ 2 & 0 \\ 0 & 3 \\ 1 & 2 \end{bmatrix} = \begin{bmatrix} X \\ Y \end{bmatrix}
$$

then

$$
AB = \begin{bmatrix} I_2 & O_{23} \\ O_{22} & P \end{bmatrix} \begin{bmatrix} X \\ Y \end{bmatrix} = \begin{bmatrix} (I_2X + O_{23}Y) \\ (O_{22}X + PY) \end{bmatrix} = \begin{bmatrix} X \\ PY \end{bmatrix}
$$

$$
PY = \begin{bmatrix} 2 & 0 & 3 \\ 0 & 3 & 1 \end{bmatrix} \begin{bmatrix} 2 & 0 \\ 0 & 3 \\ 1 & 2 \end{bmatrix} = \begin{bmatrix} 7 & 6 \\ 1 & 11 \end{bmatrix}
$$

$$
\Rightarrow AB = \begin{bmatrix} X \\ PY \end{bmatrix} = \begin{bmatrix} 1 & 2 \\ 3 & 4 \\ 7 & 6 \\ 1 & 11 \end{bmatrix}
$$

This is somewhat faster than the direct evaluation of

$$AB = \begin{bmatrix} 1 & 0 & 0 & 0 & 0 \\ 0 & 1 & 0 & 0 & 0 \\ 0 & 0 & 2 & 0 & 3 \\ 0 & 0 & 0 & 3 & 1 \end{bmatrix} \begin{bmatrix} 1 & 2 \\ 3 & 4 \\ 2 & 0 \\ 0 & 3 \\ 1 & 2 \end{bmatrix} = \begin{bmatrix} 1 & 2 \\ 3 & 4 \\ 7 & 6 \\ 1 & 11 \end{bmatrix}$$

The partitioning of the matrices in a matrix multiplication must be such that all matrix products are defined.

## Additional Examples

Example 3.21
Find the complete set of (2×2) matrices that commute with $P = \begin{bmatrix} 0 & 1 \\ 0 & 0 \end{bmatrix}$.

Let the general (2×2) matrix be $A = \begin{bmatrix} a & b \\ c & d \end{bmatrix}$. Then

$$AP = \begin{bmatrix} a & b \\ c & d \end{bmatrix} \begin{bmatrix} 0 & 1 \\ 0 & 0 \end{bmatrix} = \begin{bmatrix} 0 & a \\ 0 & c \end{bmatrix}$$

and

$$PA = \begin{bmatrix} 0 & 1 \\ 0 & 0 \end{bmatrix} \begin{bmatrix} a & b \\ c & d \end{bmatrix} = \begin{bmatrix} c & d \\ 0 & 0 \end{bmatrix}$$

$AP = PA$ if and only if $c = 0$ and $d = a$.

Therefore the complete set of (2×2) matrices that commute with $P = \begin{bmatrix} 0 & 1 \\ 0 & 0 \end{bmatrix}$ is

$A = \left\{ \begin{bmatrix} a & b \\ 0 & a \end{bmatrix} \right\}$, where $a$ and $b$ are any real numbers.

Example 3.22

For the matrix $A = \begin{bmatrix} I & X \\ O & -I \end{bmatrix}$, where $X, I$ and $O$ are all square matrices of the same size $(k \times k)$, find an expression for any natural number power of $A$, $A^n$.

$$A^2 = \begin{bmatrix} I_k & X \\ O_{kk} & -I_k \end{bmatrix} \begin{bmatrix} I_k & X \\ O_{kk} & -I_k \end{bmatrix} = \begin{bmatrix} I_k & O_{kk} \\ O_{kk} & I_k \end{bmatrix} = I_{(2k)}$$

$$\Rightarrow \quad A^3 = A^2A = IA = A, \quad A^4 = A^2A^2 = I, \quad A^5 = A^4A$$
$$= A, \quad \text{etc.}$$

Therefore

$$A^n = \begin{cases} I & (n \text{ even}) \\ A & (n \text{ odd}) \end{cases}$$

The topic of adjacency matrices for directed graphs (textbook page 46) will be explored in an assignment.

Example 3.23 (Textbook, exercises 2.2, page 48, question 11)

Given that $A \begin{bmatrix} 1 \\ -1 \\ 2 \end{bmatrix} = O = A \begin{bmatrix} 2 \\ 0 \\ 3 \end{bmatrix}$ and that

$X_0 = \begin{bmatrix} 2 \\ -1 \\ 3 \end{bmatrix}$ is a solution to $AX = B$,

find a two-parameter family of solutions to $AX = B$.

The homogeneous system $AX = O$ has at least a two-parameter family of solutions

$$X_h = \begin{bmatrix} 1 \\ -1 \\ 2 \end{bmatrix} s + \begin{bmatrix} 2 \\ 0 \\ 3 \end{bmatrix} t, \qquad (s, t \in \mathbb{R})$$

We have a particular solution to the inhomogeneous system $AX = B$,

$$X_p = X_0 = \begin{bmatrix} 2 \\ -1 \\ 3 \end{bmatrix}$$

Therefore a two-parameter family of solutions to $AX = B$ is

$$X = X_p + X_h = \begin{bmatrix} 2 \\ -1 \\ 3 \end{bmatrix} + \begin{bmatrix} 1 \\ -1 \\ 2 \end{bmatrix} s + \begin{bmatrix} 2 \\ 0 \\ 3 \end{bmatrix} t, \qquad (s, t \in \mathbb{R})$$

## 3.6 Matrix Inverses

For $(n \times n)$ matrices $A$, $B$, if
$$AB = BA = I$$
then $B = A^{-1}$ is the **inverse matrix** of $A$.

A matrix that possesses an inverse is **invertible**. A non-invertible matrix is **singular**.

Example 3.24

Show that $B = \begin{bmatrix} -5 & 2 \\ 3 & -1 \end{bmatrix}$ is the inverse of $A = \begin{bmatrix} 1 & 2 \\ 3 & 5 \end{bmatrix}$.

$$AB = \begin{bmatrix} 1 & 2 \\ 3 & 5 \end{bmatrix} \begin{bmatrix} -5 & 2 \\ 3 & -1 \end{bmatrix} = \begin{bmatrix} 1 & 0 \\ 0 & 1 \end{bmatrix} = I$$

and

$$BA = \begin{bmatrix} -5 & 2 \\ 3 & -1 \end{bmatrix} \begin{bmatrix} 1 & 2 \\ 3 & 5 \end{bmatrix} = \begin{bmatrix} 1 & 0 \\ 0 & 1 \end{bmatrix} = I$$

Therefore $B$ is the inverse matrix of $A$.

If the inverse to a matrix $A$ exists, then it is unique.

Proof:
Suppose that matrices $B$ and $C$ are both inverses of $A$. Then
$AB = BA = I$ and $AC = CA = I$.
$\Rightarrow C = IC = (BA)C = B(AC) = BI = B$

The inverse matrix, if it exists, is therefore unique.

From Example 3.17, $A = \begin{bmatrix} -1 & k \\ 0 & 1 \end{bmatrix}$ is its own inverse for all values of the real number $k$:

$$A^2 = \begin{bmatrix} -1 & k \\ 0 & 1 \end{bmatrix} \begin{bmatrix} -1 & k \\ 0 & 1 \end{bmatrix} = \begin{bmatrix} 1 & 0 \\ 0 & 1 \end{bmatrix} = I.$$

Therefore in this case $A^{-1} = A$, even though $A$ is not $\pm I$.

The uniqueness of the inverse allows us to check just one of $A^{-1}A = I$ or $AA^{-1} = I$ .

## Inverse of a (2×2) Matrix

The **adjugate** (or **adjoint**) of a (2×2) matrix $A = \begin{bmatrix} a & b \\ c & d \end{bmatrix}$ is

$$\text{adj}(A) = \begin{bmatrix} d & -b \\ -c & a \end{bmatrix}.$$

$$A \text{ adj}(A) = \begin{bmatrix} a & b \\ c & d \end{bmatrix}\begin{bmatrix} d & -b \\ -c & a \end{bmatrix} = \begin{bmatrix} ad-bc & 0 \\ 0 & ad-bc \end{bmatrix} = (ad-bc)I$$

The **determinant** of $A$ is defined to be $\det A = ad - bc$.

For all (2×2) matrices such that $\det A \neq 0$, it is clear that

$$A^{-1} = \frac{\text{adj}(A)}{\det(A)} = \frac{1}{ad-bc}\begin{bmatrix} d & -b \\ -c & a \end{bmatrix}$$

A matrix whose determinant is zero is singular (has no inverse).

<u>Example 3.25</u>

$$A = \begin{bmatrix} 1 & 2 \\ 3 & 5 \end{bmatrix} \quad \Rightarrow \quad \det(A) = 1 \times 5 - 2 \times 3 = -1 \quad \text{and} \quad \text{adj}(A)$$
$$= \begin{bmatrix} 5 & -2 \\ -3 & 1 \end{bmatrix}$$

$$\Rightarrow \quad A^{-1} = \frac{\text{adj}(A)}{\det(A)} = -\begin{bmatrix} 5 & -2 \\ -3 & 1 \end{bmatrix} = +\begin{bmatrix} -5 & 2 \\ 3 & -1 \end{bmatrix}$$

<u>Example 3.26</u>

$$A = \begin{bmatrix} -1 & k \\ 0 & 1 \end{bmatrix} \quad \Rightarrow \quad \det(A) = 1 \times (-1) - (-k) \times 0$$
$$= -1 \quad \text{and} \quad \text{adj}(A) = \begin{bmatrix} 1 & -k \\ 0 & -1 \end{bmatrix}$$

$$\Rightarrow \quad A^{-1} = \frac{\text{adj}(A)}{\det(A)} = -\begin{bmatrix} 1 & -k \\ 0 & -1 \end{bmatrix} = +\begin{bmatrix} -1 & k \\ 0 & 1 \end{bmatrix} = A$$

In a square linear system ($n$ equations in $n$ unknowns), if the coefficient matrix $A$ has rank $n$, then it is invertible and

$AX = B \Rightarrow A^{-1}AX = A^{-1}B \Rightarrow IX = A^{-1}B \Rightarrow$
the solution to the linear system is
$$X = A^{-1}B$$
and the solution is [necessarily] unique.

If rank $A < n$, then $A^{-1}$ does not exist and the system is either inconsistent or has infinitely many solutions, but not a unique solution.

Example 3.27

Solve the linear system
$$\begin{array}{rcl} 3x_1 + 2x_2 & = & 10 \\ 5x_1 + 4x_2 & = & 8 \end{array}$$

$A = \begin{bmatrix} 3 & 2 \\ 5 & 4 \end{bmatrix} \Rightarrow \text{adj}(A) = \begin{bmatrix} 4 & -2 \\ -5 & 3 \end{bmatrix}$ and $det(A) = 12 - 10 = 2$

$\Rightarrow A^{-1} = \dfrac{\text{adj}(A)}{det(A)} = \dfrac{1}{2}\begin{bmatrix} 4 & -2 \\ -5 & 3 \end{bmatrix}$

The unique solution to the linear system is

$X = A^{-1}B = \dfrac{1}{2}\begin{bmatrix} 4 & -2 \\ -5 & 3 \end{bmatrix}\begin{bmatrix} 10 \\ 8 \end{bmatrix} = \dfrac{1}{2}\begin{bmatrix} (40-16) \\ (-50+24) \end{bmatrix} = \begin{bmatrix} 12 \\ -13 \end{bmatrix}$

Check by substituting the solution into the left side of the linear system:

$$\begin{array}{rclclcl} 3x_1 + 2x_2 & = & 3 \times 12 + 2 \times (-13) & = & 36 - 26 & = & 10 \\ 5x_1 + 4x_2 & = & 5 \times 12 + 4 \times (-13) & = & 60 - 52 & = & 8 \end{array}$$

## 3.7 Matrix Inversion by Gaussian Elimination

If matrix $A$ is invertible, then the reduced row-echelon form of $[\,A\ I\,]$ is $[\,I\ A^{-1}\,]$.
The details are on page 54 of the textbook.

Example 3.28 (Textbook, page 59, exercises 2.3, question 2(c), modified)

Find the inverse of $A = \begin{bmatrix} 1 & 0 & -1 \\ 3 & 2 & 0 \\ -1 & -1 & 0 \end{bmatrix}$ and hence solve the linear system

$$x - z = 1$$
$$3x + 2y = -3$$
$$-x - y = 2$$

$$[A|I] = \begin{bmatrix} 1 & 0 & -1 & 1 & 0 & 0 \\ 3 & 2 & 0 & 0 & 1 & 0 \\ -1 & -1 & 0 & 0 & 0 & 1 \end{bmatrix}$$

$\xrightarrow{R_2 - 3R_1 R_3 + R_1} \begin{bmatrix} 1 & 0 & -1 & 1 & 0 & 0 \\ 0 & 2 & 3 & -3 & 1 & 0 \\ 0 & -1 & -1 & 1 & 0 & 1 \end{bmatrix}$

$\xrightarrow{R_2 \times \frac{1}{2}} \begin{bmatrix} 1 & 0 & -1 & 1 & 0 & 0 \\ 0 & 1 & \frac{3}{2} & -\frac{3}{2} & \frac{1}{2} & 0 \\ 0 & -1 & -1 & 1 & 0 & 1 \end{bmatrix}$

$\xrightarrow{R_3 + R_2} \begin{bmatrix} 1 & 0 & -1 & 1 & 0 & 0 \\ 0 & 1 & \frac{3}{2} & -\frac{3}{2} & \frac{1}{2} & 0 \\ 0 & 0 & \frac{1}{2} & -\frac{1}{2} & \frac{1}{2} & 1 \end{bmatrix}$

$\xrightarrow{R_3 \times 2} \begin{bmatrix} 1 & 0 & -1 & 1 & 0 & 0 \\ 0 & 1 & \frac{3}{2} & -\frac{3}{2} & \frac{1}{2} & 0 \\ 0 & 0 & 1 & -1 & 1 & 2 \end{bmatrix}$

$\xrightarrow{R_1 + R_3 R_2 - \frac{3}{2}R_3} \begin{bmatrix} 1 & 0 & 0 & 0 & 1 & 2 \\ 0 & 1 & 0 & 0 & -1 & -3 \\ 0 & 0 & 1 & -1 & 1 & 2 \end{bmatrix}$

Therefore $A^{-1} = \begin{bmatrix} 0 & 1 & 2 \\ 0 & -1 & -3 \\ -1 & 1 & 2 \end{bmatrix}$

One can easily verify that $AA^{-1} = A^{-1}A = I$.

The linear system is $AX = B$, where $B = [\,1\ -3\ 2\,]^\mathsf{T}$

$$\Rightarrow \quad X \;=\; A^{-1}B \;=\; \begin{bmatrix} 0 & 1 & 2 \\ 0 & -1 & -3 \\ -1 & 1 & 2 \end{bmatrix} \begin{bmatrix} 1 \\ -3 \\ 2 \end{bmatrix} \;=\; \begin{bmatrix} 1 \\ -3 \\ 0 \end{bmatrix}$$

Therefore the unique solution is $(x, y, z) = \underline{(1, -3, 0)}$

Check of the solution:

$$AX \;=\; \begin{bmatrix} 1 & 0 & -1 \\ 3 & 2 & 0 \\ -1 & -1 & 0 \end{bmatrix} \begin{bmatrix} 1 \\ -3 \\ 0 \end{bmatrix} \;=\; \begin{bmatrix} 1 \\ -3 \\ 2 \end{bmatrix} \;=\; B$$

Example 3.29

Find the inverse of $A \;=\; \begin{bmatrix} 1 & -3 & 1 & -1 \\ 0 & 1 & 2 & 0 \\ 0 & 0 & 1 & 4 \\ 0 & 0 & 0 & 1 \end{bmatrix}$

$$[A|I] \;=\; \left[\begin{array}{cccc|cccc} 1 & -3 & 1 & -1 & 1 & 0 & 0 & 0 \\ 0 & 1 & 2 & 0 & 0 & 1 & 0 & 0 \\ 0 & 0 & 1 & 4 & 0 & 0 & 1 & 0 \\ 0 & 0 & 0 & 1 & 0 & 0 & 0 & 1 \end{array}\right]$$

$$\xrightarrow{R_1 + 3R_2} \left[\begin{array}{cccc|cccc} 1 & 0 & 7 & -1 & 1 & 3 & 0 & 0 \\ 0 & 1 & 2 & 0 & 0 & 1 & 0 & 0 \\ 0 & 0 & 1 & 4 & 0 & 0 & 1 & 0 \\ 0 & 0 & 0 & 1 & 0 & 0 & 0 & 1 \end{array}\right]$$

$$\xrightarrow{R_1 - 7R_3\, R_2 - 2R_3} \left[\begin{array}{cccc|cccc} 1 & 0 & 0 & -29 & 1 & 3 & -7 & 0 \\ 0 & 1 & 0 & -8 & 0 & 1 & -2 & 0 \\ 0 & 0 & 1 & 4 & 0 & 0 & 1 & 0 \\ 0 & 0 & 0 & 1 & 0 & 0 & 0 & 1 \end{array}\right]$$

$$\xrightarrow[R_2 + 8R_4]{R_1 + 29R_4\, R_3 - 4R_4} \left[\begin{array}{cccc|cccc} 1 & 0 & 0 & 0 & 1 & 3 & -7 & 29 \\ 0 & 1 & 0 & 0 & 0 & 1 & -2 & 8 \\ 0 & 0 & 1 & 0 & 0 & 0 & 1 & -4 \\ 0 & 0 & 0 & 1 & 0 & 0 & 0 & 1 \end{array}\right]$$

Therefore

$$A^{-1} = \begin{bmatrix} 1 & 3 & -7 & 29 \\ 0 & 1 & -2 & 8 \\ 0 & 0 & 1 & -4 \\ 0 & 0 & 0 & 1 \end{bmatrix}$$

One can easily verify that $AA^{-1} = A^{-1}A = I$.

The following statements for an $(n \times n)$ matrix $A$ are either all true or all false:

1)  $A^{-1}$ exists (that is, $A$ is invertible).
2)  The reduced row-echelon form of $A$ is $I_n$.
3)  $AX = O$ has only the trivial solution $X = O$.
4)  $AX = B$ has a unique solution for every choice of $B$.

Example 3.30   (textbook, page 59, exercises 2.3, question 4(a))

Given $A^{-1} = \begin{bmatrix} 1 & -1 & 3 \\ 2 & 0 & 5 \\ -1 & 1 & 0 \end{bmatrix}$, solve the system of equations

$$AX = \begin{bmatrix} 1 \\ -1 \\ 3 \end{bmatrix}.$$

The system has a unique solution because $A^{-1}$ exists.

$$X = A^{-1} \begin{bmatrix} 1 \\ -1 \\ 3 \end{bmatrix} = \begin{bmatrix} 1 & -1 & 3 \\ 2 & 0 & 5 \\ -1 & 1 & 0 \end{bmatrix} \begin{bmatrix} 1 \\ -1 \\ 3 \end{bmatrix} = \begin{bmatrix} 11 \\ 17 \\ -2 \end{bmatrix}$$

Example 3.31   (textbook, page 61, exercises 2.3, question 24 modified)

Show that if the block matrix $M = \begin{bmatrix} A & X \\ O & B \end{bmatrix}$ is invertible, then the matrices $A$ and $B$ are invertible **and** find $M^{-1}$. Hence find $M^{-1}$ when $M = \begin{bmatrix} 1 & 2 & 0 & 0 \\ 0 & 2 & 0 & 0 \\ 0 & 0 & 1 & 0 \\ 0 & 0 & 0 & -1 \end{bmatrix}$.

Let $M^{-1} = \begin{bmatrix} C & Y \\ Z & D \end{bmatrix}$, where $C$ and $D$ are the same size as $A$ and $B$ respectively.

$MM^{-1} = \begin{bmatrix} A & X \\ 0 & B \end{bmatrix}\begin{bmatrix} C & Y \\ Z & D \end{bmatrix} = \begin{bmatrix} AC + XZ & AY + XD \\ BZ & BD \end{bmatrix} = \begin{bmatrix} I & 0 \\ 0 & I \end{bmatrix}$

$BD = I \Rightarrow D = B^{-1}$.

$B$ is invertible and $BZ = 0 \Rightarrow Z = 0$.

$\Rightarrow AC + XZ = AC + 0 = I \Rightarrow C = A^{-1}$.

Therefore if $M$ is invertible then both $A$ and $B$ are invertible.

$AY + XD = AY + XB^{-1} = 0 \Rightarrow AY = -XB^{-1}$

$\Rightarrow Y = -A^{-1}XB^{-1}$

$\therefore \quad M^{-1} = \begin{bmatrix} A & X \\ 0 & B \end{bmatrix}^{-1} = \underline{\begin{bmatrix} A^{-1} & -A^{-1}XB^{-1} \\ 0 & B^{-1} \end{bmatrix}}$

Note that it follows from this result that, for any constant $x$ and non-zero constants $a$, $b$,

$$\begin{bmatrix} a & x \\ 0 & b \end{bmatrix}^{-1} = \begin{bmatrix} \frac{1}{a} & \frac{-x}{ab} \\ 0 & \frac{1}{b} \end{bmatrix}.$$

$A = \begin{bmatrix} 1 & 2 \\ 0 & 2 \end{bmatrix} \Rightarrow A^{-1} = \begin{bmatrix} 1 & -1 \\ 0 & \frac{1}{2} \end{bmatrix}$, $B = \begin{bmatrix} 1 & 0 \\ 0 & -1 \end{bmatrix} \Rightarrow B^{-1}$

$= \begin{bmatrix} 1 & 0 \\ 0 & -1 \end{bmatrix}$

and $X = 0 \Rightarrow -A^{-1}XB^{-1} = 0$

$\Rightarrow M^{-1} = \begin{bmatrix} 1 & -1 & 0 & 0 \\ 0 & \frac{1}{2} & 0 & 0 \\ 0 & 0 & 1 & 0 \\ 0 & 0 & 0 & -1 \end{bmatrix}$

Example 3.32

Given that $A = \begin{bmatrix} 3 & -1 \\ 0 & -2 \end{bmatrix}$,

(a) Verify that $A^2 - A - 6I = 0$; and

(b) Hence find $A^{-1}$.

(a) $A^2 = \begin{bmatrix} 3 & -1 \\ 0 & -2 \end{bmatrix}\begin{bmatrix} 3 & -1 \\ 0 & -2 \end{bmatrix} = \begin{bmatrix} 9 & -1 \\ 0 & 4 \end{bmatrix}$

$$\Rightarrow \quad A^2 - A - 6I = \begin{bmatrix} 9 & -1 \\ 0 & 4 \end{bmatrix} - \begin{bmatrix} 3 & -1 \\ 0 & -2 \end{bmatrix} - \begin{bmatrix} 6 & 0 \\ 0 & 6 \end{bmatrix} = \begin{bmatrix} 0 & 0 \\ 0 & 0 \end{bmatrix} =$$

$$0$$

(b) $A^2 - A - 6I = 0 \quad \Rightarrow \quad A^{-1}(A^2 - A - 6I) = 0$

$\Rightarrow \quad A^{-1}AA - A^{-1}A - 6A^{-1}I = 0 \quad \Rightarrow \quad A - I - 6A^{-1} = 0$

$$\Rightarrow \quad A^{-1} = \frac{1}{6}(A - I) = \frac{1}{6}\begin{bmatrix} 3-1 & -1 \\ 0 & -2-1 \end{bmatrix} = \frac{1}{6}\begin{bmatrix} 2 & -1 \\ 0 & -3 \end{bmatrix}$$

Check:

$$A^{-1} = \begin{bmatrix} 3 & -1 \\ 0 & -2 \end{bmatrix}^{-1} = \frac{1}{3\times(-2)-(-1)\times0}\begin{bmatrix} -2 & 1 \\ 0 & 3 \end{bmatrix} = \frac{1}{6}\begin{bmatrix} +2 & -1 \\ 0 & -3 \end{bmatrix}$$

## Problems

3.1 If $A = \begin{bmatrix} 0 & 11 \\ 31 & 12 \\ -12 & 10 \end{bmatrix}$, and $B = \begin{bmatrix} 26 & 15 \\ 4 & 3 \\ 12 & 11 \end{bmatrix}$

Find $A + B, A - B, A \times B^T$

3.2 Write down the transpose of the following matrices:

$A = \begin{bmatrix} 11 & 12 \\ 33 & 14 \end{bmatrix}$, $B = \begin{bmatrix} 19 \\ -16 \\ -21 \\ -2 \end{bmatrix}$, $C = \begin{bmatrix} 21 & 13 & -6 \\ -4 & 10 & 5 \end{bmatrix}$

3.3 Find the transpose of $A = \begin{bmatrix} 20 & 5 & -2 \\ -5 & 10 & -1 \\ 2 & 1 & 30 \end{bmatrix}$

3.4 Find the matrix products $AB$ and $BA$ where

$A = \begin{bmatrix} 12 & 0 \\ -22 & 11 \\ 42 & 3 \end{bmatrix}$ and $B = \begin{bmatrix} 21 & 11 \\ 1 & -12 \end{bmatrix}.$

94

3.5 Find $A^2$, where $A = \begin{bmatrix} 12 & 21 \\ 13 & 42 \end{bmatrix}$

3.6 Find $A^3$, where $A = \begin{bmatrix} 12 & 21 \\ 13 & 42 \end{bmatrix}$

3.7 Find basic solutions of $AX = O$, where
$$A = \begin{bmatrix} 1 & 12 & 10 & 1 & 10 \\ 12 & 4 & 1 & 1 & -20 \\ 23 & 36 & 1 & 2 & -12 \\ 10 & 20 & 1 & 20 & -33 \end{bmatrix}$$

3.8 If $A = \begin{bmatrix} 1 & 2 \\ 3 & 5 \end{bmatrix}$ Find $A^{-1}$.

# Chapter 4

## Matrix Operations

A matrix, in general sense, represents a collection of information stored or arranged in an orderly fashion. The mathematical concept of a matrix refers to a set of numbers, variables or functions ordered in rows and columns. Such a set that can be defined as a distinct entity, the matrix, and it can be manipulated according to some basic mathematical rules.

### 4.1 Matrices: Basic Concepts

A matrix with 9 elements is given by

$$[A] = \begin{bmatrix} a_{11} & a_{12} & a_{13} \\ a_{21} & a_{22} & a_{23} \\ a_{31} & a_{32} & a_{33} \end{bmatrix} = \begin{bmatrix} 3 & 5 & 2 \\ -7 & 4 & 6 \\ 9 & 1 & 8 \end{bmatrix}$$

Matrix [A] has 3 rows and 3 columns. Each element of matrix [A] can be referred to by its row and column number. For example

$a_{23} = 6$

A computer monitor with 800 horizontal pixels and 600 vertical pixels can be viewed as a matrix of 600 rows and 800 columns.

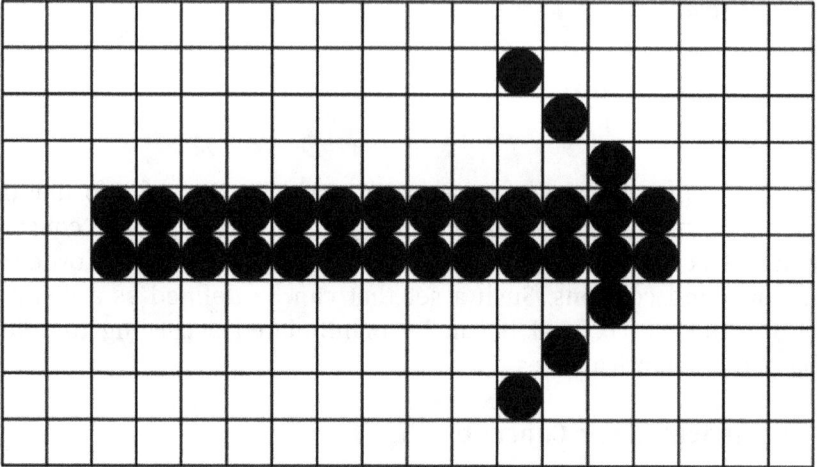
Figure 4.1 Create an image, each pixel is filled with an appropriate color.

### 4.1.1 Order of a Matrix

The order of a matrix is defined in terms of its number of rows and columns.

Order of a matrix = No. of rows × No. of columns

Matrix [A], therefore, is a matrix of order 3×3.

### COLUMN MATRIX

A matrix with only one column is called a column matrix or column vector.

$$\begin{bmatrix} 4 \\ 6 \\ -3 \end{bmatrix}$$

### ROW MATRIX
A matrix with only one row is called a row matrix or row vector

[3  5  -6 ]

### SQUARE MATRIX
A matrix having the same number of rows and columns is called a square matrix

$$\begin{bmatrix} 2 & 4 & 7 \\ -5 & 3 & 4 \\ 2 & -4 & 9 \end{bmatrix}$$

## RECTANGULAR MATRIX

A matrix having unequal number of rows and columns is called a rectangular matrix.

$$\begin{bmatrix} 5 & -3 & 7 & 1 \\ -2 & 9 & 2 & 8 \\ 5 & 4 & 1 & 13 \end{bmatrix}$$

## REAL MATRIX

A matrix with all real elements is called a real matrix

## PRINCIPAL DIAGONAL and TRACE OF A MATRIX

In a square matrix, the diagonal containing the elements $a_{11}, a_{22}, a_{33}, a_{44} \ldots, a_{nn}$ is called the principal or main diagonal.

The sum of all elements in the principal diagonal is called the trace of the matrix.
The principal diagonal of the matrix

$$\begin{bmatrix} 2 & 4 & 7 \\ -5 & 3 & 4 \\ 2 & -4 & 9 \end{bmatrix}$$

The trace of the matrix is 2+3+9 = 14.

## UNIT MATRIX

A square matrix in which all elements of the principal diagonal are equal to 1 while all other elements are zero is called the unit matrix.

$$\begin{bmatrix} 1 & 0 & 0 \\ 0 & 1 & 0 \\ 0 & 0 & 1 \end{bmatrix}$$

## ZERO Or NULL MATRIX

A matrix whose elements are all equal to zero is called the null or zero matrix

$$\begin{bmatrix} 0 & 0 & 0 \\ 0 & 0 & 0 \\ 0 & 0 & 0 \end{bmatrix}$$

## DIAGONAL MATRIX

If all elements except the elements of the principal diagonal of a square matrix are zero , the matrix is called a diagonal matrix

$$\begin{bmatrix} 2 & 0 & 0 \\ 0 & 3 & 0 \\ 0 & 0 & 9 \end{bmatrix}$$

## RANK OF A MATRIX

The maximum number of linearly independent rows of a matrix [A] is called the rank of [A] and is denoted by
Rank [A].

For a system of linear equations, a unique solution exists if the number of independent equations is at least equal to the number of unknowns

In the following system of linear equations.

$2x - 4y + 5z = 36$ ......... (1)
$-3x + 5y + 7z = 7$ ......... (2)
$5x + 3y - 8z = -31$ ......... (3)

all three equations are linearly independent. Therefor, if we form the augmented matrix [A] for the system where

$$[A] = \begin{bmatrix} 2 & -4 & 5 & 36 \\ -3 & 5 & 7 & 7 \\ 5 & 3 & -8 & -31 \end{bmatrix}$$

The rank of [A] will be 3

Consider the following linear systems with 2 independent equations.

$$2x - 4y + 5z = 36 \dots\dots (1)$$
$$-3x + 5y + 7z = 7 \dots\dots (2)$$
$$-x + y + 12z = 43 \dots\dots\dots (3)$$

In the above set, Eqn. (3) can be generated by adding Eqn. (1) to Eqn. (2). Therefore, Eqn. (3) is a dependent equation.

Therefor, if we form the augmented matrix [A] for the system where

$$[A] = \begin{bmatrix} 2 & -4 & 5 & 36 \\ -3 & 5 & 7 & 7 \\ -1 & 1 & 12 & 43 \end{bmatrix}$$

The rank of [A] will be 2.

## Equality of Matrices

Two matrices are equal if all corresponding elements are equal.

$[A] = [B]$ if $a_{ij} = b_{ij}$ for all $i$ and $j$

$$[A] = \begin{bmatrix} 2 & 4 & 3 \\ 9 & 5 & 1 \\ 3 & 7 & 8 \end{bmatrix} \qquad [B] = \begin{bmatrix} 2 & 4 & 3 \\ 9 & 5 & 1 \\ 3 & 7 & 8 \end{bmatrix}$$

## Addition and Subtraction

Two matrices can be added (subtracted) by adding (subtracting) the corresponding elements of the two matrices.

$$[C] = [A] + [B] = [B] + [A]$$
$$c_{ij} = a_{ij} + b_{ij}$$

Matrices [A], [B] and [C] must have the sameorder .

$$[A] = \begin{bmatrix} a_{11} & a_{12} & a_{13} \\ a_{21} & a_{22} & a_{23} \\ a_{31} & a_{32} & a_{33} \end{bmatrix}$$

$$[B] = \begin{bmatrix} b_{11} & b_{12} & b_{13} \\ b_{21} & b_{22} & b_{23} \\ b_{31} & b_{32} & b_{33} \end{bmatrix}$$

$$[C] = \begin{bmatrix} a_{11} + b_{11} & a_{12} + b_{12} & a_{13} + b_{13} \\ a_{21} + b_{21} & a_{22} + b_{22} & a_{23} + b_{23} \\ a_{31} + b_{31} & a_{32} + b_{32} & a_{33} + b_{33} \end{bmatrix}$$

## Multiplication by a scalar

If a matrix is multiplied by a scalar k, each element of the matrix is multiplied by k.

$$k[A] = \begin{bmatrix} ka_{11} & ka_{12} & ka_{13} \\ ka_{21} & ka_{22} & ka_{23} \\ ka_{31} & ka_{32} & ka_{33} \end{bmatrix}$$

## Matrix multiplication

Two matrices can be multiplied together provided they are compatible with respect to their orders. The number of columns in the first matrix [A] must be equal to the number of rows in the second matrix [B]. The resulting matrix [C] will have the same number of rows as [A] and the same number of columns as [B].

$$[A] = \begin{bmatrix} a_{11} a_{12} a_{13} \\ a_{21} a_{22} a_{23} \end{bmatrix} \qquad [B] = \begin{bmatrix} b_{11} & b_{12} \\ b_{21} & b_{22} \\ b_{31} & b_{32} \end{bmatrix}$$

$$[C]=[A][B]= \begin{bmatrix} a_{11} a_{12} a_{13} \\ a_{21} a_{22} a_{23} \end{bmatrix} \begin{bmatrix} b_{11} & b_{12} \\ b_{21} & b_{22} \\ b_{31} & b_{32} \end{bmatrix}$$

$$[C] = \begin{bmatrix} a_{11}b_{11} + a_{12}b_{21} + a_{13}b_{31} & a_{11}b_{12} + a_{12}b_{22} + a_{13}b_{32} \\ a_{21}b_{11} + a_{22}b_{21} + a_{23}b_{31} & a_{21}b_{12} + a_{22}b_{22} + a_{23}b_{32} \end{bmatrix}$$

$$Cij = \sum_{k=1}^{m} a_{ik} b_{kj}$$

Where m is the number of columns in [A] and the number of rows in [B]

**Example 4.1:**

$$[A] = \begin{bmatrix} 2 & 3 & 1 \\ 5 & 7 & 4 \end{bmatrix} \qquad [B] = \begin{bmatrix} 43 \\ 14 \\ 56 \end{bmatrix}$$

$$[C] = \begin{bmatrix} 2 \times 2 + 3 \times 1 + 1 \times 5 & 2 \times 3 + 3 \times 4 + 1 \times 6 \\ 5 \times 2 + 7 \times 1 + 4 \times 5 & 5 \times 3 + 7 \times 4 + 4 \times 6 \end{bmatrix}$$

$$[C] = \begin{bmatrix} 12 & 24 \\ 37 & 67 \end{bmatrix}$$

Try the following multiplication:

$$[A] = \begin{bmatrix} 2 & 1 & 4 \\ 1 & 3 & 2 \\ 4 & -2 & 5 \end{bmatrix} \qquad [B] = \begin{bmatrix} 43 \\ 12 \\ 51 \end{bmatrix}$$

$$[C]=[A][B]= \begin{bmatrix} 29 & 12 \\ 17 & 11 \\ 39 & 13 \end{bmatrix}$$

**Transpose of a Matrix**

The transpose $[A]^T$ of an $m \times n$ matrix $[A]$ is the $n \times m$ matrix obtained by interchanging the rows and columns of [A].

$$[A] = \begin{bmatrix} a_{11} & a_{12} & a_{13} \\ a_{21} & a_{22} & a_{23} \\ a_{31} & a_{32} & a_{33} \end{bmatrix} = \begin{bmatrix} 4 & 5 & 2 \\ -3 & 1 & 7 \\ 2 & 9 & 6 \end{bmatrix}$$

$$[A]^{T} = \begin{bmatrix} a_{11} & a_{12} & a_{13} \\ a_{21} & a_{22} & a_{23} \\ a_{31} & a_{32} & a_{33} \end{bmatrix} = \begin{bmatrix} 4 & -3 & 2 \\ 5 & 1 & 9 \\ 2 & 7 & 6 \end{bmatrix}$$

**Transpose of a sum**
$([A]+[B])^{T} = [A]^{T}+[B]^{T}$
**Transpose of a product**
$([A][B])^{T} = [B]^{T}[A]^{T}$

Numerical example of the product rule

$$[A] = \begin{bmatrix} 2 & 3 \\ 0 & 4 \\ 5 & 1 \end{bmatrix} \quad [B] = \begin{bmatrix} 4 & 3 & 0 & 1 \\ 2 & 1 & 5 & 3 \end{bmatrix}$$

$$([A][B])^{T} = \begin{bmatrix} 14 & 8 & 22 \\ 9 & 4 & 16 \\ 15 & 20 & 5 \\ 11 & 12 & 8 \end{bmatrix}$$

$[B]^{T}[A]^{T} = ?$

**Symmetric Matrices**
A matrix [A] is said to be symmetric if $a_{ij}=a_{ij}$ for all i and j
$[A]=[A]^{T}$

**Example 4.2:**

$$[A] = \begin{bmatrix} 4 & 3 & 2 \\ 3 & 5 & 7 \\ 2 & 7 & 0 \end{bmatrix}$$

**DETERMINANT OF A MATRIX**

Why determinants?

In some forms of solutions for systems of linear equations, determinants appear as denominators in a routine manner.

In a system with 3 unknowns, the determinant may appear in the solution in the following way.

$$x = \frac{D_x}{D} \quad y = \frac{D_y}{D} \quad z = \frac{D_z}{D}$$

$$[D] = \begin{vmatrix} a_{11} & a_{12} & a_{13} \\ a_{21} & a_{22} & a_{23} \\ a_{31} & a_{32} & a_{33} \end{vmatrix}$$

$$[D] = \begin{vmatrix} a_{11} & a_{12} & a_{13} \\ a_{21} & a_{22} & a_{23} \\ a_{31} & a_{32} & a_{33} \end{vmatrix}$$

$$[D] = \begin{vmatrix} a_{11} & a_{12} & a_{13} \\ a_{21} & a_{22} & a_{23} \\ a_{31} & a_{32} & a_{33} \end{vmatrix}$$

$$[D] = \begin{vmatrix} a_{11} & a_{12} & a_{13} \\ a_{21} & a_{22} & a_{23} \\ a_{31} & a_{32} & a_{33} \end{vmatrix}$$

$$= a_{11} \begin{vmatrix} a_{22} & a_{23} \\ a_{32} & a_{33} \end{vmatrix} - a_{21} \begin{vmatrix} a_{12} & a_{13} \\ a_{32} & a_{33} \end{vmatrix} + a_{31} \begin{vmatrix} a_{12} & a_{13} \\ a_{22} & a_{23} \end{vmatrix}$$

$$= a_{11}(a_{22}a_{33} - a_{23}a_{32}) - a_{21}(a_{12}a_{33} - a_{13}a_{32}) + a_{31}(a_{12}a_{23} - a_{13}a_{22})$$

$$D = \begin{vmatrix} 2 & -3 & 4 \\ 1 & 4 & -2 \\ 3 & 5 & 6 \end{vmatrix}$$

$$= 2 \begin{vmatrix} 4 & -2 \\ 5 & 6 \end{vmatrix} - 1 \begin{vmatrix} -3 & 4 \\ 5 & 6 \end{vmatrix} + 3 \begin{vmatrix} -3 & 4 \\ 4 & -2 \end{vmatrix}$$

$$= 2(4 \times 6 + 2 \times 5) - 1(-3 \times 6 - 4 \times 5) + 3(3 \times 2 - 4 \times 4)$$

$$= 68 + 38 - 30$$

Find the determinant

$$D = \begin{vmatrix} 3 & 1 & 4 \\ 6 & 2 & 1 \\ 7 & 0 & -5 \end{vmatrix}$$

## Important Properties of Determinants

1. The value of a determinant is not altered if its rows are written as columns in the same order.

$$\begin{vmatrix} 3 & 1 & 4 \\ 6 & 2 & 1 \\ 7 & 0 & -5 \end{vmatrix} = \begin{vmatrix} 3 & 1 & 4 \\ 6 & 2 & 1 \\ 7 & 0 & -5 \end{vmatrix}$$

2. If any two rows (or two columns) of a determinant are interchanged, the value of the determinant is multiplied by -1.

$$\begin{vmatrix} 3 & 1 & 4 \\ 6 & 2 & 1 \\ 7 & 0 & -5 \end{vmatrix} = \begin{vmatrix} 6 & 2 & 1 \\ 3 & 1 & 4 \\ 7 & 0 & -5 \end{vmatrix}$$

3. A common factor of all elements of any row (or column) can be placed before the determinant.

$$\begin{vmatrix} 3 & 8 & 1 \\ 5 & 4 & 2 \\ 1 & 12 & -3 \end{vmatrix} = \begin{vmatrix} 3 & 4 \times 2 & 1 \\ 5 & 4 \times 1 & 2 \\ 1 & 4 \times 3 & -3 \end{vmatrix} = 4 \begin{vmatrix} 3 & 2 & 1 \\ 5 & 1 & 2 \\ 1 & 3 & -3 \end{vmatrix}$$

4. If the corresponding elements of two rows (or columns) of a determinant are proportional, the value of the determinant is zero.

$$\begin{vmatrix} 3 & 2 & 5 \\ 6 & 4 & 10 \\ 2 & 7 & 8 \end{vmatrix} = 0$$

Meaning: Row 2 (Row 1) is linearly dependent on Row 1 (Row 2). Therefore, the linear system with three unknowns does not have a unique solution.

5. The value of a determinant remains unaltered if the elements of one row (or column) are altered by adding to the many constant multiple of the corresponding elements in any other row (or column).

$$\begin{vmatrix} 3 & 1 & 4 \\ 6 & 2 & 1 \\ 7 & 0 & -5 \end{vmatrix} = \begin{vmatrix} 3 + 2 \times 6 & 1 + 2 \times 2 & 4 + 2 \times 1 \\ 6 & 2 & 1 \\ 7 & 0 & -5 \end{vmatrix}$$

6. If each element of a row (or a column) of a determinant can be expressed as a sum of two, the determinant can be written as the sum of two determinants.

$$\begin{vmatrix} 3 & 1 & 4 \\ 6 & 2 & 1 \\ 7 & 0 & -5 \end{vmatrix} = \begin{vmatrix} -1 + 4 & 1 & 4 \\ 3 + 3 & 2 & 1 \\ 5 + 2 & 0 & -5 \end{vmatrix}$$

$$= \begin{vmatrix} -1 & 1 & 4 \\ 3 & 2 & 1 \\ 5 & 0 & -5 \end{vmatrix} + \begin{vmatrix} 4 & 1 & 4 \\ 3 & 2 & 1 \\ 2 & 0 & -5 \end{vmatrix}$$

$$= -49$$

7. Determinant of a product of matrices

$$D([A][B]) = D[A]D[B]$$

$$[A] = \begin{bmatrix} 2 & 3 & 4 \\ 1 & -1 & 3 \\ 4 & 2 & 1 \end{bmatrix} [B] = \begin{bmatrix} 1 & 2 & 3 \\ 4 & -6 & 5 \\ 3 & 1 & 4 \end{bmatrix}$$

$$[C] = [A][B]$$

$$[C] = \begin{bmatrix} 26 & -10 & 37 \\ 6 & 11 & 10 \\ 15 & -3 & 26 \end{bmatrix}$$

$$D[C] = D([A][B]) = 1505$$
$$D[43]$$
$$D[A]D[B] = 43 \times 35 = 1505 \text{ and } D[B] = 35$$

Matrix transformation

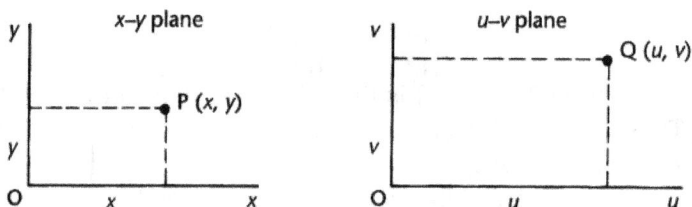

Figure 4.2

If for every point $Q$ $(u, v)$ Q (u, v) in the $u - v$ plane there is a corresponding point $P$ $(x, y)$ in the $x - y$ plane, then there is a relationship between the two sets of coordinates. In the simple case of scaling the coordinate where

$u = ax$ and $v = by$

we have a linear transformation, and we can combine these in matrix form

$$\begin{pmatrix} u \\ v \end{pmatrix} = \begin{pmatrix} a & 0 \\ 0 & b \end{pmatrix} \begin{pmatrix} x \\ y \end{pmatrix}$$

The matrix $\begin{pmatrix} a & 0 \\ 0 & b \end{pmatrix}$ then provides the transformation between the vector $\begin{pmatrix} x \\ y \end{pmatrix}$ in one set of coordinates and the vector $\begin{pmatrix} u \\ v \end{pmatrix}$ in the other set of coordinates.

Similarly, if we solve the two equations for $x$ and $y$, we have

$x = \dfrac{1}{a}u$ and $y = \dfrac{1}{a}v$

$$\begin{pmatrix} u \\ v \end{pmatrix} = \begin{pmatrix} \dfrac{1}{a} & 0 \\ 0 & \dfrac{1}{b} \end{pmatrix} \begin{pmatrix} x \\ y \end{pmatrix}$$

which allows us to transform back from the u-v plane coordinates to the $x - y$ plane coordinates.
Now for an example.

**Example 4.3:**

If $X = \begin{pmatrix} x \\ y \end{pmatrix} = \begin{pmatrix} 2 \\ 1 \end{pmatrix}$ with the transformation $T = \begin{pmatrix} -2 & 0 \\ 2 & 1 \end{pmatrix}$ determine $U = \begin{pmatrix} u \\ v \end{pmatrix} = TX$ and show the positions on the $x - y$ and $u - v$ planes.
In this case

$$\begin{pmatrix} u \\ v \end{pmatrix} = \begin{pmatrix} -2 & 0 \\ 2 & 1 \end{pmatrix} \begin{pmatrix} 2 \\ 1 \end{pmatrix} = \begin{pmatrix} -4 \\ 5 \end{pmatrix}$$

transforms into

Figure 4.3

If $T$ is non-singular and $U = TX$ then $X = T^{1}U$ and since

$T = \begin{pmatrix} -1/2 & 0 \\ 1 & 1 \end{pmatrix}$ then $T^{-1} = \dots\dots\dots\dots$

$$T^{-1} = \begin{pmatrix} -1/2 & 0 \\ 1 & 1 \end{pmatrix}$$

There are several ways of finding the inverse of a matrix. One method is as follows.

$$T = \begin{pmatrix} -1/2 & 0 \\ 1 & 1 \end{pmatrix}$$

$$\begin{pmatrix} -2 & 0 & 1 & 0 \\ 2 & 1 & 0 & 1 \end{pmatrix} \sim \begin{pmatrix} -2 & 0 & 1 & 0 \\ 2 & 1 & 0 & 1 \end{pmatrix} \sim \begin{pmatrix} 1 & 0 & -1/2 & 0 \\ 0 & 1 & 1 & 1 \end{pmatrix}$$

107

$$T^{-1} = \begin{pmatrix} -1/2 & 0 \\ 1 & 1 \end{pmatrix}$$

So we have $U = TX \therefore X = T - 1U$

$$\begin{pmatrix} x \\ y \end{pmatrix} = \begin{pmatrix} -\dfrac{1}{2} & 0 \\ 1 & 1 \end{pmatrix} \begin{pmatrix} u \\ v \end{pmatrix}$$

Hence a vector $\begin{pmatrix} 1 \\ 4 \end{pmatrix}$ in the $u - v$ transforms into $\begin{pmatrix} x \\ y \end{pmatrix}$ in the $x - y$ plane where $\begin{pmatrix} x \\ y \end{pmatrix} = \ldots\ldots\ldots\ldots$

$$\begin{pmatrix} x \\ y \end{pmatrix} = \begin{pmatrix} -1/2 \\ 5 \end{pmatrix}$$

$$\begin{pmatrix} x \\ y \end{pmatrix} = \begin{pmatrix} -\dfrac{1}{2} & 0 \\ 1 & 1 \end{pmatrix} \begin{pmatrix} 1 \\ 4 \end{pmatrix} = \begin{pmatrix} -1/2 \\ 5 \end{pmatrix}$$

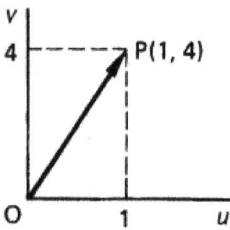

transforms into

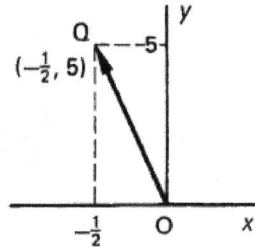

Figure 4.4

## Rotation of axes

A more interesting case occurs with a degree of rotation between the two sets of coordinate axes.

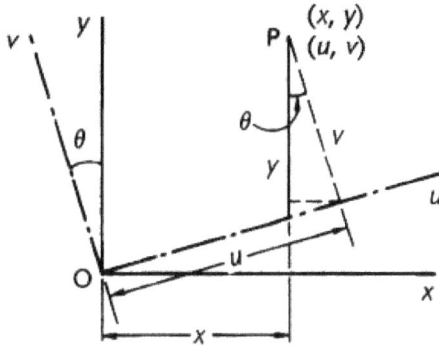

Let P be the point $(x, y)$ in the $x - y$ plane and the point $(u, v)$ in the $u - v$ plane.

Let $\theta$ be the angle of rotation between the two systems. From the diagram we can see that

$$x = u\cos\theta - v\sin\theta \atop y = u\sin\theta + v\cos\theta \Big\}$$

In matrix form, this becomes

$$\binom{x}{y} = \begin{pmatrix} \cos\theta & -\sin\theta \\ \sin\theta & \cos\theta \end{pmatrix}\binom{u}{v}$$

which enables us to transform from the $u - v$ plane coordinates to the corresponding $x - y$ plane coordinates.

If we solve equations (1) for $u$ and $v$, we have

$x\sin\theta = u\sin\theta\cos\theta - v\sin^2\theta$

$y\cos\theta = u\sin\theta\cos\theta + v\cos^2\theta$

$y\cos\theta - x\sin\theta = v(\cos^2\theta + \sin^2\theta) = v$

Also

$x\cos\theta = u\cos^2\theta - v\sin\theta\cos\theta$

$y\sin\theta = u\sin^2\theta + v\sin\theta\cos\theta$

$x\cos\theta + y\sin\theta = u(\cos^2\theta + \sin^2\theta) = u$

So

$u = x\cos\theta + y\sin\theta$

$v = -x\sin\theta + y\cos\theta$

and written in matrix form, this is

$$\binom{u}{v} = \begin{pmatrix} \cos\theta & \sin\theta \\ -\sin\theta & \cos\theta \end{pmatrix}\binom{x}{y}$$

So we have

$$\binom{x}{y} = \begin{pmatrix} \cos\theta & -\sin\theta \\ \sin\theta & \cos\theta \end{pmatrix}\binom{u}{v}$$

$$\binom{u}{v} = \begin{pmatrix} \cos\theta & \sin\theta \\ -\sin\theta & \cos\theta \end{pmatrix}\binom{x}{y}$$

109

and
$i.e. X = TU$ and $U = T^1X$
where $T$ is the matrix of transformation and the equations provide a linear transformation between the two sets of coordinates.

## Example 4.4:
If the $u - v$ plane axes rotate through 30° in an anticlockwise manner from the $x - y$ plane axes, determine the $(u, v)$ coordinates of a point whose $(x, y)$ coordinates are $x = 2, y = 3$ in the $x - y$ plane. This is a straightforward application of the results above.
So

$$\binom{u}{v} = \cdots \cdots \cdots \cdots \cdots \cdots$$

$$\binom{u}{v} = \begin{pmatrix} \sqrt{3} + 3/2 \\ -1 + 3\sqrt{3}/2 \end{pmatrix} = \begin{pmatrix} 3.23 \\ 1.60 \end{pmatrix}$$

because

$$\binom{u}{v} = \begin{pmatrix} \cos\theta & \sin\theta \\ -\sin\theta & \cos\theta \end{pmatrix} \binom{2}{3} \quad \begin{matrix} \cos\theta = \sqrt{3}/2 \\ \sin\theta = 1/2 \end{matrix}$$

$$= \begin{pmatrix} \sqrt{3}/2 & 1/2 \\ -1/2 & \sqrt{3}/2 \end{pmatrix} \binom{2}{3}$$

$$= \begin{pmatrix} \sqrt{3} + 3/2 \\ -1 + 3\sqrt{3}/2 \end{pmatrix} = \begin{pmatrix} 3.23 \\ 1.60 \end{pmatrix}$$

## 4.1.2 Complex conjugate matrix

If $A = \begin{bmatrix} 1 & 2 + 4j \\ 3 & 5j \\ 1 - j & 2 + 6j \end{bmatrix}$ then $\bar{A} = \begin{bmatrix} 1 & 2 - 4j \\ 3 & -5j \\ 1 + j & 2 - 6j \end{bmatrix}$

$\bar{A}$ Is the complex conjugate matrix of A
Unit matrix

$$I = \begin{bmatrix} 1 & 0 & 0 \\ 0 & 1 & 0 \\ 0 & 0 & 1 \end{bmatrix} \quad A.I = I.A = A$$

### 4.1.3 Determinant

If A is a square matrix $A = \begin{bmatrix} a_{11} & a_{12} & a_{13} \\ a_{21} & a_{22} & a_{23} \\ a_{31} & a_{32} & a_{33} \end{bmatrix}$    $\Delta A = det\ A =$

$\begin{vmatrix} a_{11} & a_{12} & a_{13} \\ a_{21} & a_{22} & a_{23} \\ a_{31} & a_{32} & a_{33} \end{vmatrix}$

a)  Minor: given any element of a j k of Δ we associate anew determinant of order (n-1) obtained be removing an elements of the (j th) row and (hth) colon, this is called the minor of a j k.

Example

$\begin{vmatrix} 2 & -1 & 1 & 3 \\ 3 & 2 & \boxed{5} & 0 \\ 1 & 0 & -2 & 2 \\ 4 & -2 & 3 & 1 \end{vmatrix} = \begin{vmatrix} 2 & -1 & 3 \\ 1 & 0 & 2 \\ 4 & -2 & 1 \end{vmatrix}$ the minor of element 5 is

b)  Cofactor :if we multiply the minor of a j k by (- 1)$^{j+k}$ the result is called the cofactor of a j k is determined by A j k.

Example

$(-1)^{2+3} = - \begin{vmatrix} 2 & -1 & 3 \\ 1 & 0 & 2 \\ 4 & -2 & 1 \end{vmatrix}$

Now    $detA = \Delta A = \sum_{k=1}^{n} \underbrace{ajk}_{element} \quad \underbrace{Ajk}_{cofactor}$

Example

If $A = \begin{bmatrix} 1 & 3 & 2 \\ 4 & 5 & 7 \\ 2 & 4 & 8 \end{bmatrix}$        $\Delta A = 1 \begin{vmatrix} 5 & 7 \\ 4 & 8 \end{vmatrix} - 3 \begin{vmatrix} 4 & 7 \\ 2 & 8 \end{vmatrix} +$

$2 \begin{vmatrix} 4 & 5 \\ 2 & 4 \end{vmatrix} = -30$

$= 1 \times (40 - 28) - 3 \times (32 - 14) + 2 \times (16 - 10) = -30$

### 4.1.4 Adjoint of matrix

If A is a square matrx defined as:

$$A = \begin{bmatrix} a_{11} & a_{12} & a_{13} \\ a_{21} & a_{22} & a_{23} \\ a_{31} & a_{32} & a_{33} \end{bmatrix}$$

we can form a new matrix C of cofactors of the cofactors

$$C = \begin{bmatrix} A_{11} & A_{12} & A_{13} \\ A_{21} & A_{22} & A_{23} \\ A_{31} & A_{32} & A_{33} \end{bmatrix} \text{ the adjoint of A (adj)= C}^\text{T}$$

Example

$$A = \begin{bmatrix} 2 & 3 & 5 \\ 4 & 1 & 6 \\ 1 & 4 & 0 \end{bmatrix} \rightarrow C =$$

$$\begin{bmatrix} \overset{A_{11}}{+\begin{vmatrix} 1 & 6 \\ 4 & 0 \end{vmatrix}} & \overset{A_{12}}{-\begin{vmatrix} 4 & 6 \\ 1 & 0 \end{vmatrix}} & \overset{A_{13}}{+\begin{vmatrix} 4 & 1 \\ 1 & 4 \end{vmatrix}} \\ \overset{A_{21}}{-\begin{vmatrix} 3 & 5 \\ 4 & 0 \end{vmatrix}} & \overset{A_{22}}{+\begin{vmatrix} 2 & 5 \\ 1 & 0 \end{vmatrix}} & \overset{A_{23}}{-\begin{vmatrix} 2 & 3 \\ 1 & 4 \end{vmatrix}} \\ \overset{A_{31}}{+\begin{vmatrix} 3 & 5 \\ 1 & 6 \end{vmatrix}} & \overset{A_{32}}{-\begin{vmatrix} 2 & 5 \\ 4 & 6 \end{vmatrix}} & \overset{A_{33}}{+\begin{vmatrix} 2 & 3 \\ 4 & 1 \end{vmatrix}} \end{bmatrix} \quad C = \begin{bmatrix} -24 & 6 & 15 \\ 20 & -5 & -5 \\ 13 & 8 & -10 \end{bmatrix}$$

$$adjA = C \overset{traspos}{\overset{\rightarrow}{^T}} = \begin{bmatrix} -24 & 20 & 13 \\ 6 & -5 & 8 \\ 15 & -5 & -10 \end{bmatrix}$$

## 4.1.5 Matrix inversion

$A^{-1} = \frac{adjA}{\det A}$ if is non-singular matrix $[\det A \neq 0]$

For the above example $\det A = 45$

$$A^{-1} = \frac{1}{45} \begin{bmatrix} -24 & 20 & 13 \\ 6 & -5 & 8 \\ 15 & -5 & -10 \end{bmatrix}$$

$A^{-1}A = I$

**Example 4.5** : if $A = \begin{bmatrix} 2 & 1 & 1 \\ 1 & 3 & 0 \\ 2 & 1 & 0 \end{bmatrix}$ and $B = \begin{bmatrix} 1 & 3 & 1 \\ 2 & 4 & 3 \\ 1 & 1 & 0 \end{bmatrix}$

prove that

$(AB)^{-1} = A^{-1}B^{-1}$

   1.   USchut format for inversion

$A \in \mathbb{R}^{2*2} = \begin{bmatrix} a_{11} & a_{12} \\ a_{21} & a_{22} \end{bmatrix}$

$A^{-1} = \dfrac{1}{\det \begin{bmatrix} a_{22} & -a_{12} \\ -a_{21} & a_{11} \end{bmatrix}}$

**Example 4.6:**

$A = \begin{bmatrix} 1 & 2 \\ 3 & 4 \end{bmatrix} \Rightarrow \det A = 4 - 6 = -2$

$$A^{-1} = \frac{1}{-2}\begin{bmatrix} 4 & -2 \\ -3 & 1 \end{bmatrix}$$

2. For diagonal matrix of order n*n

$$A = \begin{bmatrix} a_{11} & 0 & 0 & \cdots & 0 \\ 0 & a_{22} & 0 & \cdots & 0 \\ 0 & 0 & a_{33} & \cdots & 0 \\ \vdots & \vdots & \vdots & \ddots & 0 \\ 0 & \cdots & \cdots & \cdots & a_{nn} \end{bmatrix}$$

$$A^{-1} = \begin{bmatrix} \frac{1}{a_{11}} & 0 & 0 & \cdots & 0 \\ 0 & \frac{1}{a_{22}} & 0 & \cdots & 0 \\ 0 & 0 & \frac{1}{a_{33}} & \cdots & 0 \\ \vdots & \vdots & \vdots & \ddots & 0 \\ 0 & \cdots & \cdots & \cdots & \frac{1}{a_{nn}} \end{bmatrix}$$

Inverse of Product
$$(AC)^{-1} = C^{-1}A^{-1}$$
In general, $(APQR)^{-1} = R^{-1}Q^{-1}P^{-1}A^{-1}$

## 4.2 System of linear equations

A set of equations leving the form

$$a_{11}x_1 + a_{12}x_2 + \cdots \ldots \ldots \ldots \ldots \ldots + a_{1n}x_n = r_1$$
$$a_{21}x_1 + a_{22}x_2 + \cdots \ldots \ldots \ldots \ldots \ldots + a_{2n}x_n = r_2$$
$$\vdots$$
$$\vdots$$

$$a_{m1}x_1 + a_{n2}x_2 + \cdots \ldots \ldots \ldots \ldots \ldots \ldots + a_{mn}x_n = r_m$$
Is called a system of (m) liner equations in the (n) unknowns (x₁, x₂, ....xₙ) if (r₁, r₂,... rₙ,) are zero the system is called homogeneous.
Any set of numbers (x₁, x₂, ....xₙ) satity , the above equation is called the solution in matrix form.

$$\underbrace{\begin{bmatrix} a_{11} & a_{12} & a_{13} & \cdots & a_{1n} \\ a_{21} & a_{22} & a_{23} & \cdots & a_{2n} \\ \vdots & \vdots & \vdots & \cdots & \cdots \\ \vdots & \vdots & \vdots & \cdots & \cdots \\ a_{m1} & a_{m2} & a_{m3} & \cdots & a_{mn} \end{bmatrix}}_{A} \underbrace{\begin{bmatrix} x_1 \\ x_2 \\ x_3 \\ \vdots \\ x_n \end{bmatrix}}_{X} = \underbrace{\begin{bmatrix} r_1 \\ r_2 \\ r_3 \\ \vdots \\ r_n \end{bmatrix}}_{R} \qquad X = A^{-1}R$$

## 4.3 Elementary operation on matrices

The following operators can be carried on matrices
1- Inter changing of columns or arrows
2- Multiplication of row (columns) by a non- zero number
3- Addition of a numbers multiplied arrow (column) to another row(column) .

Solution of linear equation by Elementary operation (Gaussian – elimination)

**Example 4.7:**
Solve the following operations

$$x_1 + 2x_2 + x_3 = 2$$

$$3x_1 + x_2 - 2x_3 = 1$$

$$4x_1 - 3x_2 - x_3 = 3$$

$$2x_1 + 4x_2 + 2x_3 = 4$$

**Solution**

$$\begin{bmatrix} 1 & 2 & 1 & | & 2 \\ 3 & 1 & -2 & | & 1 \\ 4 & -3 & -1 & | & 3 \\ 2 & 4 & 2 & | & 4 \end{bmatrix} \begin{matrix} \\ \leftarrow R_2 = R_2 - 3R_1 \\ \leftarrow R_3 = R_3 - 4R_1 \\ \leftarrow R_4 = R_4 - 2R_1 \end{matrix}$$

$$AH$$

$$\begin{bmatrix} 1 & 2 & 1 & | & 2 \\ 0 & -5 & -5 & | & -5 \\ 0 & -11 & -5 & | & -5 \\ 0 & 0 & 0 & | & 0 \end{bmatrix} \leftarrow R_2 = \frac{R_2}{-5} \Rightarrow \begin{bmatrix} 1 & 2 & 1 & | & 2 \\ 0 & 1 & 1 & | & 1 \\ 0 & -11 & -5 & | & -5 \\ 0 & 0 & 0 & | & 0 \end{bmatrix}$$

$$\Rightarrow \begin{bmatrix} 1 & 2 & 1 & | & 2 \\ 0 & 1 & 1 & | & 1 \\ 0 & -11 & -5 & | & -5 \\ 0 & 0 & 0 & | & 0 \end{bmatrix} \begin{matrix} \leftarrow R_1 = R_1 - 2R_2 \\ \\ \leftarrow R_3 = R_3 + 11R_2 \end{matrix}$$

$$\Rightarrow \begin{bmatrix} 1 & 0 & -1 & | & 0 \\ 0 & 1 & 1 & | & 1 \\ 0 & 0 & 6 & | & 6 \\ 0 & 0 & 0 & | & 0 \end{bmatrix} \leftarrow R_3 = \frac{R_3}{6}$$

114

$$\Rightarrow \begin{bmatrix} 1 & 0 & -1 & | & 0 \\ 0 & 1 & 1 & | & 1 \\ 0 & 0 & 1 & | & 1 \\ 0 & 0 & 0 & | & 0 \end{bmatrix} \begin{matrix} \leftarrow R_1 = R_1 + R_3 \\ \leftarrow R_2 = R_2 - R_3 \end{matrix} \Rightarrow \begin{bmatrix} 1 & 0 & 0 & | & 1 \\ 0 & 1 & 0 & | & 0 \\ 0 & 0 & 1 & | & 1 \\ 0 & 0 & 0 & | & 0 \end{bmatrix}$$

$CK$

There is unique solution and that is :

$$x = \begin{bmatrix} 1 \\ 0 \\ 1 \end{bmatrix}$$

$i.e\, x_1 = 1, x_2 = 0, x_3 = 1$

**Example 4.8:**

Solve by R.E.O

$$x_1 + 2x_2 - 3x_3 = 3$$

$$2x_1 - x_2 - x_3 = 11$$

$$3x_1 + 2x_2 + x_3 = -5$$

$$\Rightarrow \begin{bmatrix} 1 & 2 & -3 & | & 3 \\ 2 & -1 & -1 & | & 11 \\ 3 & 2 & 1 & | & -5 \end{bmatrix} \begin{matrix} \leftarrow R_2 = R_2 - 2R_1 \\ \leftarrow R_3 = R_3 - 3R_1 \end{matrix}$$

$AH$

$$\begin{bmatrix} 1 & 2 & -3 & | & 3 \\ 0 & -5 & 5 & | & 5 \\ 0 & -4 & 10 & | & -14 \end{bmatrix} \leftarrow R_2 = \frac{R_2}{-5}$$

$$\Rightarrow \begin{bmatrix} 1 & 2 & -3 & | & 3 \\ 0 & 1 & -1 & | & -1 \\ 0 & -4 & 10 & | & -14 \end{bmatrix} \begin{matrix} \leftarrow R_1 = R_1 - 2R_2 \\ \\ \leftarrow R_3 = R_3 + 4R_2 \end{matrix}$$

$$\begin{bmatrix} 1 & 0 & -1 & | & 5 \\ 0 & 1 & -1 & | & -1 \\ 0 & 0 & 6 & | & -18 \end{bmatrix} \leftarrow R_3 = \frac{R_3}{6}$$

$$\Rightarrow \begin{bmatrix} 1 & 0 & -1 & | & 5 \\ 0 & 1 & -1 & | & -1 \\ 0 & 0 & 1 & | & -3 \end{bmatrix} \begin{matrix} \leftarrow R_1 = R_1 + R_3 \\ \leftarrow R_2 = R_2 + R_3 \end{matrix}$$

115

$$\Rightarrow \begin{bmatrix} 1 & 0 & 0 & | & 2 \\ 0 & 1 & 0 & | & -4 \\ 0 & 0 & 1 & | & -3 \end{bmatrix}$$
$$\phantom{xx} C \phantom{xxxx} K$$

The solution is : $\quad x = \begin{bmatrix} x_1 \\ x_2 \\ x_3 \end{bmatrix} = \begin{bmatrix} 2 \\ -4 \\ -3 \end{bmatrix}$ and it unique

### Note 1
In both above examples we have

$[A/H] \sim [C/K]$

C is the row equivalent Canonical from of A.

A has rank 3: 1st three rows of C have non-zero component. First non-zero component in each of three rows of C is fourth zeros in the rest of that column .the four the row of $[C/K]$ (if any component of zeros $[A/H]$ has rank 3

i.e $r = 3 \leftarrow$rows of $[A/H]$

$n = 3 \leftarrow$No of unknowns

Here $n = r$

Therefore, if we have $A.x = H$ .(m) linear equations in (n) unknowns and the rank of $[A]$and $[A/H]$is $r = n$ there is a unique solution. This system of equation is called <u>consistent</u>

### Example 4.9
Solve

$$x_1 + 2x_2 - 3x_3 - 4x_4 = 6$$

$$x_1 + 3x_2 + x_3 - 2x_4 = 4$$

$$2x_1 + 5x_2 - 2x_3 - 5x_4 = 10$$

**Solution**: the argument of matrix is

$$\begin{bmatrix} 1 & 2 & -3 & -4 & | & 6 \\ 1 & 3 & 1 & -2 & | & 4 \\ 2 & 5 & -2 & -5 & | & 10 \end{bmatrix} \begin{matrix} \\ \leftarrow R_2 = R_2 - R_1 \\ \leftarrow R_3 = R_3 - 2R_1 \end{matrix}$$

$$\Rightarrow \begin{bmatrix} 1 & 2 & -3 & -4 & | & 6 \\ 0 & 1 & 4 & 2 & | & -2 \\ 0 & 1 & 4 & 3 & | & -2 \end{bmatrix} \begin{matrix} \leftarrow R_1 = R_1 - 2R_2 \\ \\ \leftarrow R_3 = R_3 - R_2 \end{matrix}$$

116

$$\begin{bmatrix} 1 & 0 & -11 & -8 & | & 10 \\ 0 & 1 & 4 & 2 & | & -2 \\ 0 & 0 & 0 & 1 & | & 0 \end{bmatrix} \begin{matrix} \leftarrow R_1 = R_1 + 8R_3 \\ \leftarrow R_2 = R_2 - R_3 \end{matrix} \Rightarrow \begin{bmatrix} 1 & 0 & -110 & | & 10 \\ 0 & 1 & 4 & 0 & | & -2 \\ 0 & 0 & 0 & 1 & | & 0 \end{bmatrix}$$

CK

Let $x_1 = a$     arbitrary unknowns

$$X_1 - 11X_3 = 10$$

let     $X_1 = a$

$$a - 11X_3 = 10 \Rightarrow X_3 = \frac{10 - a}{-11}$$

$$X_2 + 4X_3 = -2$$

$$X_2 = -2 - 4X_3$$

$$X_2 = -2 - 4\left(\frac{10-a}{-11}\right), \quad X_4 = 0$$

We have infinite set of solution C is a rows equivalent Canonical form ,A has rank 3because C has all three rows of non –zero element . $[AH]$ Has rank 3.

$r = 3 \leftarrow$ rank of $[AH]$

$n = 3 \leftarrow$ No of unknowns

$n > r$ Therefore, if the rank of $A$ and $[AH]$ $(r)$ less than n , there are $(n > r)$ infinite set of solution

**Example 4.10**: solve

$$x_1 + x_2 = 3$$

$$x_1 + 2x_2 = 5$$

$$2x_1 + x_2 = 2$$

**Solution**: The argument matrix $[A/H]$ is

$$\begin{bmatrix} 1 & 1 & |3 \\ 1 & 2 & |5 \\ 2 & 1 & |2 \end{bmatrix} \begin{matrix} \\ \leftarrow R_2 = R_2 - R_1 \\ \leftarrow R_3 - 2R_1 \end{matrix} \Rightarrow \begin{bmatrix} 1 & 1 & | & 3 \\ 0 & 1 & | & 2 \\ 0 & -1 & | & -4 \end{bmatrix} \begin{matrix} \leftarrow R_1 = R_1 - R_2 \\ \\ \leftarrow R_3 = R_3 + R_2 \end{matrix}$$

$$\Rightarrow \begin{bmatrix} 1 & 0 & | & 1 \\ 0 & 1 & | & 2 \\ 0 & 0 & | & -2 \end{bmatrix} \begin{matrix} \\ \\ \leftarrow R_3 = \dfrac{R_3}{-2} \end{matrix}$$

*AH*

$$\begin{bmatrix} 1 & 0 & |1 \\ 0 & 1 & |2 \\ 0 & 0 & |1 \end{bmatrix} \begin{matrix} x_1 = 1 \\ \Rightarrow x_2 = 2 \Rightarrow \\ 0 = 1 \end{matrix} \begin{bmatrix} 1 & 1 & |3 \\ 1 & 2 & |5 \\ 2 & 1 & |2 \end{bmatrix} \sim \begin{bmatrix} 1 & 0 & |1 \\ 0 & 1 & |2 \\ 0 & 0 & |1 \end{bmatrix}$$

*CKAHCK*

A has rank 2 because 1st two rows of C have non zero elements
$[A/H]$ has rank 3 , r=3 rank of AH  n=2 no of unknowns n<r.
therefore if n<r

The system of equation is consistent and has non solution

**Note** : on rank

1)      By E.O rank of matrix is not changed

2)      If any rows column of matrix $A \in R^{n \times n}$  has its elements equal zero than the rank of $A = n - 1$

**Exersice: Solve**

$$3x_1 + 2x_2 + x_3 = 3$$

$$2x_1 + x_2 + x_3 = 0$$

$$6x_1 + 2x_2 + 4x_3 = 6$$

4.4 Inverse of matrix by E.R.C (Gaussian elimination method)
$AB = I \Rightarrow B = A^{-1}, AX = H$

**Example 4.11**

Find the inverse of

$$A = \begin{bmatrix} 1 & 3 & 3 \\ 1 & 4 & 3 \\ 1 & 3 & 4 \end{bmatrix}$$

**Solution**

The argument of matrix $[A|H]$

$$\begin{bmatrix} 1 & 3 & 3 & | & 1 & 0 & 0 \\ 1 & 4 & 3 & | & 0 & 1 & 0 \\ 1 & 3 & 4 & | & 0 & 0 & 1 \end{bmatrix}$$

$$\begin{bmatrix} 1 & 3 & 3 & | & 1 & 0 & 0 \\ 0 & 1 & 0 & | & -1 & 1 & 0 \\ 0 & 0 & 1 & | & -1 & 0 & 1 \end{bmatrix} \begin{matrix} \\ \leftarrow R_2 = R_2 - R_1 \\ \leftarrow R_3 = R_3 - R_1 \end{matrix}$$

$$\begin{bmatrix} 1 & 0 & 3 & | & 4 & -3 & 0 \\ 0 & 1 & 0 & | & -1 & 1 & 0 \\ 0 & 0 & 1 & | & -1 & 0 & 1 \end{bmatrix} \leftarrow R_1 = R_1 - 3R_2$$

$$\begin{bmatrix} 1 & 0 & 0 & | & 7 & -3 & -3 \\ 0 & 1 & 0 & | & -1 & 1 & 0 \\ 0 & 0 & 1 & | & -1 & 0 & 1 \end{bmatrix} \leftarrow R_1 = R_1 - 3R_3$$

$IA^{-1}$

**H.w : Find $A^{-1}$**

$$A = \begin{bmatrix} 2 & 4 & 3 & 2 \\ 3 & 6 & 5 & 2 \\ 2 & 5 & 2 & -3 \\ 4 & 5 & 14 & 14 \end{bmatrix}$$

Check $AA^{-1} = I$

## 4.5 Eigen values and Eigen vectors

A scalar constant number $\lambda$ is called eigen value of the square matrix A if where exists a non –zero solution of the equation $Ax = \lambda x \rightarrow 1$

Any vector $x \neq 0$ that satisfies equation (1) is called eigen vector associated with the eigen value $\lambda$ ,from (1)

$$Ax - \lambda x = 0$$

$$(A - \lambda I)x = 0..........2$$

To have a non-zero solution of this set homogenies linear equation (2) $|A - \lambda I|$ must be equal to zero .

i.e $\quad |A - \lambda I| = 0........(3)$

Equation 3 is called the characteristic equation.

If $\lambda$ is a parameter, then write

$P(\lambda) = det[A - \lambda I]$ Which is called characteristic polynomial and $p(\lambda)$ zero C/CS equation

The following procedure can find the eigen values of eigen vectors of n order matrix

1.      To find the C/CS polynomial $p(\lambda) = det([A - \lambda I])$
2.      To find the roots of the C/CS equation $p(\lambda) = 0$.
The roots are eigen values that we require
3.      To solve the homogenous system $|A - \lambda I|x = 0$
To find n-eigen vector

**Example 4.12:** Determine the eigen value and corresponding eigen vector of the matrix

$$A = \begin{bmatrix} 4 & 1 \\ 3 & 2 \end{bmatrix}$$

$$|A - \lambda I| = 0 \Rightarrow \left\|\begin{bmatrix} 4 & 1 \\ 3 & 2 \end{bmatrix} - \begin{bmatrix} \lambda & 0 \\ 0 & \lambda \end{bmatrix}\right\| = 0$$

$$\begin{vmatrix} 4 - \lambda & 1 \\ 3 & 2 - \lambda \end{vmatrix} = 0 \Rightarrow \begin{matrix} (4 - \lambda)(2 - \lambda) - 3 = \\ (\lambda^2 - 6\lambda + 5) = 0 \\ (\lambda - 1)(\lambda - 5) = 0 \end{matrix}$$

$$\therefore \lambda = 1, \lambda = 5$$

To find the eigen vectors for $\lambda = 1$
$$Ax = \lambda x$$

$$\begin{bmatrix} 4 & 1 \\ 3 & 2 \end{bmatrix}\begin{bmatrix} x_1 \\ x_2 \end{bmatrix} = 1\begin{bmatrix} x_1 \\ x_2 \end{bmatrix}$$

$4x_1 + x_2 = x_1 \rightarrow 3x_1 + x_2 = 0$
$3x_1 + 2x_2 = x_2 \rightarrow 3x_1 + x_2 = 0$
Let $x_1 = a$   (arbitrary constant)
$x_2 = -3a$
The eigen vectors corresponding to $\lambda = 1$ is $X_1 = \begin{bmatrix} a \\ -3a \end{bmatrix}$ eigen vector
may be normalized to unit length the normalized eigen vector $\hat{X}$ is defined by $\hat{X} = \frac{X}{Length}$
For $X_i::Length = a\sqrt{1 + 3} = a\sqrt{4} = 2a$

$$\therefore \hat{X} = \frac{a}{a\sqrt{4}}\begin{bmatrix} 1 \\ -3 \end{bmatrix} = \begin{bmatrix} \frac{1}{2} \\ \frac{-3}{2} \end{bmatrix}$$

For $\lambda_2 = 5 \rightarrow AX = \lambda_2 X$

120

$$\begin{bmatrix} 4 & 1 \\ 3 & 2 \end{bmatrix}\begin{bmatrix} x_1 \\ x_2 \end{bmatrix} = 5\begin{bmatrix} x_1 \\ x_2 \end{bmatrix}$$

$$4x_1 + x_2 = 5x_1 \rightarrow x_1 - x_2 = 0$$
$$3x_1 + 2x_2 = 5x_2 \rightarrow 3x_2 - 3x_1 = 0$$

$x_1 = x_2$
Let

$$x_1 = a$$

$x_2 = a$

The eigen vector corresponding to $\lambda_2 = 5$ is $X_2 = \begin{bmatrix} a \\ a \end{bmatrix}$

$$\hat{X}_2 = \frac{a}{a\sqrt{2}}\begin{bmatrix} 1 \\ 1 \end{bmatrix} = \begin{bmatrix} \frac{1}{2} \\ \frac{1}{2} \end{bmatrix}$$

**Example 4.13:**  Find Eigen values of corresponding Eigen vector of the matrix

$$A = \begin{bmatrix} 6 & -2 & 2 \\ -2 & 5 & 0 \\ 2 & 0 & 7 \end{bmatrix}$$

$$p(\lambda) = |A - \lambda I| = 0$$

$$\begin{bmatrix} 6-\lambda & -2 & 2 \\ -2 & 5-\lambda & 0 \\ 2 & 0 & 7-\lambda \end{bmatrix} = 0$$

Simplifying we have

$$P(\lambda) = \lambda^3 - 18\lambda^2 + 99\lambda - 162 = 0$$

$$\lambda = g$$

$$(\lambda - g)(\lambda^2 - 9\lambda + 18) = 0$$

$$(\lambda - g)(\lambda - 3)(\lambda - 6) = 0$$

$$\therefore \lambda_1 = 3\lambda_2 = 6\lambda_3 = 9$$

Case  $\lambda = 3, (A - \lambda_i I)xi = 0$
$(A - \lambda_1 I)X_1 = 0$

$$\begin{bmatrix} 3 & -2 & 2 \\ -2 & 2 & 0 \\ 2 & 0 & 4 \end{bmatrix}\begin{bmatrix} x_1 \\ x_2 \\ x_3 \end{bmatrix} = 0$$

$$3x_1 - 2x_2 + 2x_3 = 0 \ldots\ldots\ldots\ldots\ldots\ldots (1)$$

$$-2x_1 + 2x_2 + 0 = 0 \ldots\ldots\ldots\ldots\ldots\ldots (2)$$

$$2x_1 + 0 + 4x_3 = 0 \ldots\ldots\ldots\ldots\ldots\ldots (3)$$

Let $X_3 = -C_1, X_1 = X_2 = 2C_1$

The eigen vectors is $X_1 = \begin{bmatrix} 2C_1 \\ 2C_1 \\ -C_1 \end{bmatrix} = C_1 \begin{bmatrix} 2 \\ 2 \\ -1 \end{bmatrix}$

Case $\lambda_2 = 6$

$(A - \lambda I)X_2 = 0$

$$\begin{bmatrix} 0 & -2 & 2 \\ -2 & -1 & 0 \\ 2 & 0 & 1 \end{bmatrix}\begin{bmatrix} x_1 \\ x_2 \\ x_3 \end{bmatrix} = 0$$

$$-2x_2 + 2x_3 = 0 \ldots\ldots\ldots\ldots\ldots (1)$$

$$-2x_1 - x_2 + 0 = 0 \ldots\ldots\ldots\ldots\ldots (2)$$

$$2x_1 + 0 + x_3 = 0 \ldots\ldots\ldots\ldots\ldots (1) \therefore Eigenvector$$

$x_1 = C_2$

Let $x_2 = -2C_2 X_2 = C_2 \begin{bmatrix} 1 \\ -2 \\ -2 \end{bmatrix}$

$x_3 = -2C_2$

Case $\lambda_3 = 9 (A - \lambda_3 I)X_3 = 0$

$$\begin{bmatrix} -2 & -2 & 2 \\ -2 & -4 & 0 \\ 2 & 0 & -2 \end{bmatrix}\begin{bmatrix} x_1 \\ x_2 \\ x_3 \end{bmatrix} = 0 , \text{ this gives}$$

$X_3 = C_3 \begin{bmatrix} -2 \\ 1 \\ -2 \end{bmatrix}$  $C_1, C_2, C_3$ are arbitrary constant

**Theorem**

Let A be a square matrix of order n the matrix $[P^{-1}AP]$ is a diagonal matrix if any of the columns of P are a set of n linearly independent eigen vector of A

The elements of this diagonal matrix are the corresponding eigen values $P^{-1}AP = D_g(\lambda_1, \lambda_i, \lambda_A)$

**Example 4.14**: find all eigen values of vectors of the matrix A check that $P^{-1}AP$ is diagonal matrix

$$A = \begin{bmatrix} 8 & -8 & 8 \\ 9 & -15 & -18 \\ -5 & 11 & 14 \end{bmatrix}$$

$$P(\lambda) = |A - \lambda I| = 0$$
$$P(\lambda) = \lambda^3 - I\lambda^2 + 12\lambda = 0$$
$$\lambda.(\lambda^2 - I\lambda + 12) = 0$$
$$\lambda(\lambda - 3)(\lambda - 4) = 0$$

$$\lambda_1 = 4, \lambda_2 = 3, \lambda_3 = 0$$

**Case:** $\lambda_1 = 4, (A - 4I)X_1 = 0$

$$\begin{bmatrix} 4 & -8 & 8 \\ 9 & -19 & -18 \\ -5 & 11 & 10 \end{bmatrix}\begin{bmatrix} x_1 \\ x_2 \\ x_3 \end{bmatrix} = \begin{bmatrix} 0 \\ 0 \\ 0 \end{bmatrix}$$

$X_1 = \begin{bmatrix} 2 \\ 0 \\ 1 \end{bmatrix}$ Similarity we can find

$$X_2 = \begin{bmatrix} 0 \\ 1 \\ 1 \end{bmatrix}, X_3 = \begin{bmatrix} 1 \\ 3 \\ 2 \end{bmatrix}$$

We can constant the matrix P

$$P = [X_1|X_2|X_3]$$

$$P = \begin{bmatrix} 2 & 0 & 1 \\ 0 & 1 & 3 \\ 1 & 1 & -2 \end{bmatrix}$$

$$P^{-1} = \begin{bmatrix} -1 & 2 & 2 \\ 3 & -5 & -6 \\ 1 & -1 & -1 \end{bmatrix}, P^{-1}AP = \begin{bmatrix} 0 & 0 & 0 \\ 0 & 3 & 0 \\ 0 & 0 & 4 \end{bmatrix}$$

## 4.6 Cayley –Hamilton Theorem

let A be a square matrix of order n where C/CS poly is
$$P(\lambda) = det(A - \lambda I) = a_0 + a_1\lambda + a_2\lambda^2 + \ldots\ldots + a_{n-1}\lambda^{n-1} + a_n\lambda^n$$
Scaler poly
Then $a_0I + a_1A + a_2A^2 + \ldots\ldots + a_{n-1}A^{n-1} + a_nA^n = 0$

**Example 4.15: let**

$$A = \begin{bmatrix} 2 & 2 & 1 \\ 1 & 3 & 1 \\ 1 & 2 & 2 \end{bmatrix}$$

$$P(\lambda) = |A - \lambda I| = 0$$

$$P(\lambda) = \begin{vmatrix} 2-\lambda & 2 & 1 \\ 1 & 3-\lambda & 1 \\ 1 & 2 & 2-\lambda \end{vmatrix} = 0$$

$$P(\lambda) = \lambda^3 - 7\lambda^2 + 11\lambda - 5 = 0C/CSeq$$

$$A^3 - 7A^2 + 11A - 5I = 0..........2$$

$$A^3 = \begin{bmatrix} 32 & 62 & 31 \\ 31 & 63 & 31 \\ 31 & 62 & 32 \end{bmatrix} = A.A.A, \quad A^2 = \begin{bmatrix} 7 & 12 & 6 \\ 6 & 13 & 6 \\ 6 & 12 & 7 \end{bmatrix} = A.A$$

i.e equ(2) becomes

$$[A^3] - 7[A^2] + 11[A] - 5\begin{bmatrix} 1 & 0 & 0 \\ 0 & 1 & 0 \\ 0 & 0 & 1 \end{bmatrix} = \begin{bmatrix} 0 & 0 & 0 \\ 0 & 0 & 0 \\ 0 & 0 & 0 \end{bmatrix}$$

**Example 4.16:** Using Cayley –Hamilton theorem to find the inverse of A where:

$$A = \begin{bmatrix} 1 & 1 & 2 \\ 3 & 1 & 1 \\ 2 & 3 & 1 \end{bmatrix}$$

**Solution**

$$P(\lambda) = |A - \lambda I| = 0$$

$$P(\lambda) = \begin{vmatrix} 1-\lambda & 1 & 2 \\ 3 & 1-\lambda & 1 \\ 2 & 3 & 1-\lambda \end{vmatrix} = 0 \rightarrow \begin{array}{l} \lambda^3 - 3\lambda^2 - 7\lambda - 11 = 0 \\ A^3 - 3A^2 - 7A - 11I = 0 \end{array}$$

To find $A^{-1}$

$$A^3 - 3A^2 - 7A = 11I$$

Multiplying both sides by $A^{-1}$

$$A^2 - 3A - 7I = 11A^{-1}$$

$$A^{-1} = \frac{1}{n}[A^2 - 3A - 7I]$$

124

$$A^{-1} = \frac{1}{n}\left\{\begin{bmatrix} 8 & 8 & 5 \\ 8 & 7 & 8 \\ 13 & 8 & 8 \end{bmatrix} - 3\begin{bmatrix} 1 & 1 & 2 \\ 3 & 1 & 1 \\ 2 & 3 & 1 \end{bmatrix} - 7\begin{bmatrix} 1 & 0 & 0 \\ 0 & 1 & 0 \\ 0 & 0 & 1 \end{bmatrix}\right\}$$

$$A^{-1} = \frac{1}{n}\begin{bmatrix} -2 & 5 & 1 \\ -1 & -3 & 5 \\ 7 & -1 & -2 \end{bmatrix}$$

Find $A^{-2}$

$$A - 3I - 7A^{-1} = 11A^{-2}$$

$$A^{-2} = \frac{1}{11}[A - 3I - 7A^{-1}]$$

**H.w**

If $A = \begin{bmatrix} 2 & 1 & 2 \\ 0 & 2 & 3 \\ 0 & 0 & 5 \end{bmatrix}$

Using Cayley-Hamilton theorem to show it to

a) $A^{-1} = \frac{1}{20}\begin{bmatrix} 10 & -5 & -1 \\ 0 & 10 & -6 \\ 0 & 0 & 4 \end{bmatrix}$ and

b) $A^4 = \begin{bmatrix} 16 & 32 & 517 \\ 0 & 16 & 609 \\ 0 & 0 & 625 \end{bmatrix}$

Example if $A = \begin{bmatrix} 1 & 0 & 0 \\ 0 & -1 & 1 \\ 2 & 0 & 2 \end{bmatrix}$

Compute

a) $P(A) = 3A^4 + 2A^2 - 4I$

b) $P(A) = A^{-3} + 4A^2$

A-C/CS $\quad C(\lambda) = det\begin{bmatrix} \lambda - 1 & 0 & 0 \\ 0 & \lambda + 1 & -1 \\ -2 & 0 & \lambda - 2 \end{bmatrix} = (\lambda - 1)(\lambda + 2)(\lambda - 2)$

$$= \lambda^3 - 2\lambda^2 - \lambda + 2 \quad \text{scale poly}$$

**C.H theorem**

$$C(A) = A^3 - 2A^2 + 2I = 0$$
$$A^3 = 2A^2 + A - 2I.A$$

$$A^4 = A.A^3 = 2A^3 + A^2 - 2A$$

$$A^4 = 4A^2 + 2A - 4I + A^2 - 2A = 5A^2 - 4I$$

$$P(A) = 3(5A^2 - 4I) + 2A^2 - 4I$$

125

$$P(A) = 17A^2 - 16I$$

B- $(A^{-3}) = (A^{-1})^3 C(A) = A^3 - 2A - A + 2I$

By C.H theorem

$$A^{-1} = -\frac{1}{2}[A^2 - 2A - I]$$

$$A^{-2} = -\frac{1}{2}[A - 2I - A^{-1} + A.A^{-1}]$$

$$A^{-3} = \frac{1}{8}[I - 2A^{-1} - A - 2I - A^{-1}]$$

$$A^{-3} = -\frac{1}{8}[-4A + 3A^2 - 3I]$$

$$A^{-3} = -\frac{1}{8}[3A^2 - 4A - 3I]$$

**Example 4.17**: If $A = \begin{bmatrix} 1 & 0 & 1 \\ -1 & 0 & -1 \\ 2 & 1 & 0 \end{bmatrix}$ compute cosh A

**Solution**

$$C(\lambda) = |\lambda - IA| = \begin{bmatrix} \lambda - 1 & 0 & 1 \\ 1 & \lambda & -1 \\ -2 & 1 & \lambda \end{bmatrix}$$

$$= \lambda^3 - \lambda^2 - \lambda$$

$$P(\lambda) = \lambda^3 - \lambda^2 - \lambda = \lambda(\lambda^2 - \lambda - 1)$$

$$\lambda_1 = 0, \lambda_2 = \frac{1 - \sqrt{5}}{2}, \frac{1 + \sqrt{5}}{2}$$

$$\cosh A = \frac{\cosh \lambda_1}{(\lambda_1 - \lambda_2)(\lambda_1 - \lambda_3)}[A - \lambda_2 I][A - \lambda_3 I]$$

$$+ \frac{\cosh \lambda_2}{(\lambda_2 - \lambda_1)(\lambda_2 - \lambda_3)}[A - \lambda_1 I][A - \lambda_3 I]$$

$$+ \frac{\cosh \lambda_3}{(\lambda_3 - \lambda_1)(\lambda_3 - \lambda_2)}[A - \lambda_1 I][A - \lambda_2 I]$$

**Exercise**: If $A = \begin{bmatrix} 1 & 0 & 2 \\ 0 & -1 & 2 \\ 2 & -2 & 0 \end{bmatrix}$

a) Show that $\sin A = \left(\dfrac{\sin 3}{3}\right) A$

b) Compute $A^{20}$

c) Compute $\cos A$

**Example 4.18**: If $A = \begin{bmatrix} 3 & -1 \\ 6 & -2 \end{bmatrix}$

**Comput : $e^A$, $\cos A$**

1- $\quad e^A = I + \dfrac{A}{1!} + \dfrac{A^2}{2!} + \dfrac{A^3}{3!} + \ldots\ldots + \dfrac{A^n}{n!} = f(I, A)$

C/CS $\quad C(\lambda) = \begin{vmatrix} \lambda - 3 & 1 \\ -6 & \lambda + 2 \end{vmatrix} = (\lambda - 3)(\lambda + 2) + 6 = \lambda^2 - \lambda$

By C.H theorem

$$A^2 - A = 0, A^2 = A, A^3 = A.A^2 = A.A = A^2 = A$$

$$A^n = A$$

$$e^A = I + \frac{A}{1!} + \frac{A}{2!} + \ldots\ldots + \frac{A^n}{n!}$$

$$e^A = I + A\left(\frac{1}{1!} + \frac{1}{2!} + \ldots\ldots + \frac{1}{n!}\right)$$

$$e^A = \begin{bmatrix} 1 & 0 \\ 0 & 1 \end{bmatrix} + \begin{bmatrix} 3 & 1 \\ 6 & -2 \end{bmatrix}(e - 1)$$

$$e^A = \begin{bmatrix} 6.83 & -1.718 \\ 10.31 & -1.718 \end{bmatrix}$$

2-

$$\cos A = I + \frac{A^2}{2!} + \frac{A^4}{4!} + \ldots\ldots$$

$$= I + A\left(\frac{1}{2!} + \frac{1}{4!} + \ldots\ldots\right)$$

$$\cos A = I + A(\cos 1 - 1)$$

For distinct eigen values we state Sylvesler's identity to compte function of eigen matrix, such as $\cos A$, $\sin A$, $\cosh A$

$$f(A) = \sum_{k=1}^{n} \left[ \frac{f(\lambda_k) \quad \overset{\overset{i=1}{i\neq k}}{\underset{i=1}{\pi}}{(A - \lambda_i I)}}{\underset{\underset{i\neq k}{\pi}}{(\lambda_k - \lambda_i)}} \right]$$

$$\lambda_1 \neq \lambda_2 \neq \lambda_3 \neq \lambda_4$$

When n is the order of eigen matrix

For example, above

$$e^A = \sum_{\substack{k=1 \\ i \neq k}}^{2} \left[ \frac{e^{\lambda_k} \prod\limits_{\substack{i=1 \\ i \neq k}}^{i=1} (A - \lambda_i I)}{\prod\limits_{\substack{2 \\ \pi}}} \right]$$

$$\frac{e^1}{(\lambda_1 - \lambda_2)} [A - \lambda_2 I] + \frac{e^2}{(\lambda_2 - \lambda_1)} [A - \lambda_1 I]$$

$$\lambda_1 = 1$$

$$\lambda_2 = 0$$

## 4.7 Singular and non-singilar matrices

Every square matrix $A$ has associated with it a number called the determinant of $A$ and denoted by $|A|$. If $|A| \neq 0$ then $A$ is called a *non-singular* matrix. Otherwise, if $|A| = 0$, then $A$ is called *singular* matrix.

**Example 4.19**

Is $A = \begin{pmatrix} 1 & 2 & 8 \\ 4 & 7 & 6 \\ 9 & 5 & 3 \end{pmatrix}$ singular or non-singular?

$$|A| = \begin{vmatrix} 1 & 2 & 8 \\ 4 & 7 & 6 \\ 9 & 5 & 3 \end{vmatrix}$$

$$= 1 \begin{vmatrix} 7 & 6 \\ 5 & 3 \end{vmatrix} - 2 \begin{vmatrix} 4 & 6 \\ 9 & 3 \end{vmatrix} + 8 \begin{vmatrix} 4 & 7 \\ 9 & 5 \end{vmatrix}$$

$$= (21 - 30) - 2(12 - 54) + 8(20 - 63)$$

$$= -9 + 84 - 344$$

$$= -269$$

Because $|A| \neq 0$ then $A$ is non-singular.

**Example 4.20**

Is $A = \begin{pmatrix} 3 & 9 & 2 \\ 1 & 5 & 6 \\ 2 & 7 & 4 \end{pmatrix}$ singular or non-singular?

A is singular

Because $|A| = 0$ then $|A|$ is singular

128

**Exercise**
Determine whether each of the following is singular or non-singular

1. $|A| = \begin{pmatrix} 4 & 5 \\ 2 & 3 \end{pmatrix}$

2. $|B| = \begin{pmatrix} 3 & -4 \\ -6 & 8 \end{pmatrix}$

3. $|C| = \begin{pmatrix} 4 & 1 & -2 \\ 1 & 7 & 3 \\ 5 & 8 & 1 \end{pmatrix}$

4. $|D| = \begin{pmatrix} 3 & 2 & 4 \\ 5 & 1 & 6 \\ 2 & 0 & 3 \end{pmatrix}$

**1 non-singular  2 singular**
**1 singular       2 non-singular**

Straightforward evaluation of the relevant determinants gives
1. $|A| = 2$   2. $|B| = 0$
3. $|C| = 0$   4. $|D| = -5$
Closely related to the notion of the singularity or otherwise of a square matrix is the notion of the rank of a general $n \times m$ matrix.
4.8 Rank of a Matrix
The rank of an $n \times m$ matrix  **A** is the largest square, non-singular sub-matrix. That is, the largest square sub-matrix whose determinant is non-zero. If $n = m$, so making A itself square, then this sub-matrix could be the matrix A itself $

**Example 4.21**

To find the rank of the matrix $A = \begin{pmatrix} 3 & 4 & 5 \\ 1 & 2 & 3 \\ 4 & 5 & 6 \end{pmatrix}$ we note that

$$|A| = \begin{vmatrix} 3 & 4 & 5 \\ 1 & 2 & 3 \\ 4 & 5 & 6 \end{vmatrix} = .................$$

Because
$$|A| = \begin{vmatrix} 3 & 4 & 5 \\ 1 & 2 & 3 \\ 4 & 5 & 6 \end{vmatrix}$$
$$= 3(12 - 15) - 4(6 - 12) + 5(5 - 8)$$
$$= -9 + 24 - 15$$
$$= 0$$
Therefore we can say that the rank of A is Not 3 **Because** $|A| = 0$ and therefore, **A** is singular .
Now try a sub-matrix of order 2

$\begin{vmatrix} 3 & 4 \\ 1 & 2 \end{vmatrix} = 6 - 4 = -2 \neq 0$. Therefore the rank of A is 2 Because the largest square, non-singular sub-matrix of A has order 2, therefore A has rank 2.

## 4.9 LU-Decomposition for Band Matrices

The matrix **H** is sparse matrix. If the matrix $H \in R^{n \times m}$ is the regular square matrix , whose main minors are different from zero, a division (**H=LU**) can be made in a simple way. **L** denotes the lower triangular matrix, with number one on the main diagonal, and **U** is the upper triangular matrix, with non-zero diagonal elements.
According to the expected form, **L** and **U** matrices are as follows:

$$L = \begin{bmatrix} 1 & 0 & 0 & \cdots & \cdots & 0 \\ l_1^2 & 1 & 0 & \cdots & \cdots & 0 \\ \vdots & l_2^2 & 1 & \ddots & \ddots & \vdots \\ l_1^{n+1} & \vdots & \ddots & 1 & \ddots & 0 \\ 0 & \ddots & \ddots & \ddots & \ddots & 0 \\ 0 & 0 & l_{n^2-n}^{n+1} & \cdots & l_{n^2-1}^2 & 1 \end{bmatrix},$$ (4.1)

$$U = \begin{bmatrix} u_1^1 & u_1^2 & \cdots & u_1^{n+1} & 0 & 0 \\ 0 & u_2^1 & u_2^2 & \cdots & & \ddots & 0 \\ 0 & 0 & u_3^1 & u_3^2 & \cdots & u_{n^2-n}^{n+1} \\ \vdots & \vdots & \ddots & \ddots & \ddots & \vdots \\ 0 & \vdots & \ddots & \ddots & u_{n^2-1}^1 & u_{n^2-1}^2 \\ 0 & 0 & \cdots & \cdots & 0 & u_{n^2}^1 \end{bmatrix}.$$ (4.2)

The members of **L** and **U** matrices can be calculated by using the following algorithm which can be written only in three parts due to the discontinuity in the intervals of the sum:

**The first part of the algorithm:** for each $m=1, ..., n^2-n$ the following can be applied

$$u_m^p = a_m^p - \sum_{i=1}^{m-1} u_{m-i}^{i+p} \cdot l_{m-i}^{i+1}, \qquad p = 1, ..., n+1$$ (4.3)

$$l_m^p = \frac{1}{u_m^1} \left( c_m^{p-1} - \sum_{i=1}^{m-1} u_{m-1}^{i+1} \cdot l_{m-i}^{i+p} \right) \qquad p = 2, ..., n+1.$$

**The second part of the algorithm:** for each $q=1, ..., n-1$ and for each $m=n^2-n+1, ..., n^2-1$ the following can be applied

130

$$u_m^p = a_m^p - \sum_{i=1}^{m-1} u_{m-i}^{i+p} \cdot l_{m-i}^{i+1}, \qquad\qquad p = 1, ..., n - q + 1 \qquad (4.4)$$

$$l_m^p = \frac{1}{u_m^1}\left( c_m^{p-1} - \sum_{i=1}^{m-1} u_{m-i}^{i+1} \cdot l_{m-i}^{i+p} \right). \qquad p = 2, ..., n - q + 1$$

**The third part of the algorithm:** the last member of the main diagonal of the **U** matrix is determined

$$u_{n^2}^1 = a_{n^2}^1 - \sum_{i=1}^{n^2-1} u_{n^2-i}^{i+1} \cdot l_{n^2-i}^{i+1}. \qquad\qquad (4.5)$$

This algorithm can be used to determine LU-decomposition of the band matrix with any number of subordinate diagonals. The advantage of this algorithm compared to *Crout's* [1] algorithm is that during programming only the elements that are next to the main diagonal have to be entered instead of all elements of the system matrix.

Instead of the system **Hf=g**, the system **LUf=g** is observed. That system is solved successively, i.e. the lower triangular system **Lz*=g** is solved first, and then the upper triangular system **Uf=z***. If the vector of unknowns **z*** has the following form

$$z^{*T} = \begin{bmatrix} z_1^* & z_2^* & z_3^* & ... & z_{n^2-1}^* & z_{n^2}^* \end{bmatrix}. \qquad\qquad (4.6)$$

**The algorithm for the system Lz*=g:**
*The first part of the algorithm:* for each $m=1, ..., n+1$ the following can be applied

$$z_m^* = g_m - \sum_{i=1}^{m-1} l_{m-i}^{i+1} \cdot z_{m-i}^*, \qquad\qquad (4.7)$$

*The second part of the algorithm:* for each $m=n+2, ..., n^2$ the following can be applied

$$z_m^* = g_m - \sum_{i=1}^{n} l_{m-i}^{i+1} \cdot z_{m-i}^*. \qquad\qquad (4.8)$$

**The algorithm for the system Uf=z*:**
*The first part of the algorithm:* for each $m=0, ..., n$ the following can be applied

$$f_{n^2-m} = \frac{1}{u_{n^2-m}^1}\left( z_{n^2-m}^* - \sum_{i=1}^{m} u_{n^2-m}^{i+1} \cdot f_{n^2-m+i} \right), \qquad\qquad (4.9)$$

*The second part of the algorithm:* for each $m=n+1, ..., n^2-1$ the following can be applied

$$f_{n^2-m} = \frac{1}{u_{n^2-m}^1}\left( z_{n^2-m}^* - \sum_{i=1}^{n} u_{n^2-m}^{i+1} \cdot f_{n^2-m+i} \right). \qquad\qquad (4.10)$$

In this way we can solve the equation (5), i.e. the values of the function $f$ in each node of previously defined mesh with the adequate boundary conditions.

The same procedure can be used to get final solutions for the plate deflection that has the following discretization form

$$w_{i+1,j} + w_{i-1,j} + w_{i,j+1} + w_{i,j-1} - 4 \cdot w_{i,j} = f_{i,j}, \qquad (4.11)$$

Equation (4.11) has the following matrix form
$$\mathbf{H \cdot w = f}. \qquad (4.12)$$

## Problems

4.1 Find the determinant
$$A = \begin{vmatrix} 3 & 11 & 4 \\ 16 & 12 & 11 \\ 7 & 10 & -6 \end{vmatrix}$$

4.2 If $X = \begin{pmatrix} x \\ y \end{pmatrix} = \begin{pmatrix} 1 \\ 2 \end{pmatrix}$ with the transformation $T = \begin{pmatrix} -1 & 1 \\ 2 & 1 \end{pmatrix}$ determine $U = \begin{pmatrix} u \\ v \end{pmatrix} = TX$ and show the positions on the $x - y$ and $u - v$ planes.

4.3 If $X = \begin{pmatrix} x \\ y \end{pmatrix} = \begin{pmatrix} 1 \\ -2 \end{pmatrix}$ with the transformation $T = \begin{pmatrix} -1 & 1 \\ -2 & -1 \end{pmatrix}$ determine $U = \begin{pmatrix} u \\ v \end{pmatrix} = TX$ and show the positions on the $x - y$ and $u - v$ planes.

4.4 if $A = \begin{bmatrix} 2 & 10 & 4 \\ 1 & 3 & 10 \\ 2 & 1 & 10 \end{bmatrix}$ and $B = \begin{bmatrix} 11 & 3 & 1 \\ 12 & 14 & 3 \\ 11 & 1 & 10 \end{bmatrix}$
prove that
$(AB)^{-1} = A^{-1}B^{-1}$
if $A = \begin{bmatrix} 2 & 3 & 4 \\ 1 & 3 & 4 \\ 2 & 3 & 4 \end{bmatrix}$ and $B = \begin{bmatrix} 2 & 3 & 1 \\ 2 & 14 & 3 \\ 2 & 1 & 10 \end{bmatrix}$
prove that
$(AB)^{-1} = A^{-1}B^{-1}$

4.5 Solve the following operations
$$-3x_1 + 2x_2 + 2x_3 = 2$$

$$3x_1 + 2x_2 - 2x_3 = 1$$

$$4x_1 - 3x_2 - 3x_3 = 3$$

4.6  Find the inverse of

$$A = \begin{bmatrix} 2 & 3 & 3 \\ 2 & 3 & 4 \\ 2 & 3 & 4 \end{bmatrix}$$

4.7 Determine the eigen value and corresponding eigen vector of the matrix

$$A = \begin{bmatrix} 2 & 5 \\ 3 & 4 \end{bmatrix}$$

4.8 Determine the eigen value and corresponding eigen vector of the matrix

$$A = \begin{bmatrix} 2 & 1 & 3 \\ 3 & -2 & 3 \\ 2 & -2 & -3 \end{bmatrix}$$

4.9 Determine the eigen value and corresponding eigen vector of the matrix

$$A = \begin{bmatrix} 3 & 4 & 1 \\ 1 & 2 & 4 \\ 4 & 3 & -2 \end{bmatrix}$$

4.10  Find all eigen values of vectors of the matrix A check that $P^{-1}AP$ is diagonal matrix

$$A = \begin{bmatrix} 2 & -2 & 8 \\ 3 & -15 & -4 \\ -2 & 2 & 3 \end{bmatrix}$$

4.11 : If $A = \begin{bmatrix} 2 & -1 \\ 3 & -2 \end{bmatrix}$  **Compute : $e^{2A}, cos\ 3A$.**

4.12 To find the rank of the matrix $A = \begin{pmatrix} 2 & 4 & 2 \\ 1 & -3 & 3 \\ 1 & 3 & 2 \end{pmatrix}$

# Chapter 5

## Optimization

An optimization problem includes finding the minimum or maximum value of a function of multiple variables. To be familiar with principles of utilization, especially from the commercial viewpoint, fundamentals of optimization in this chapter introduce linear programming and optimization problem formulation.

### 5.1 Optimization Problem

Main components of an optimization problem:

1- Decision Variables

The n variables which we want to obtain their values at the optimum point (min or max) of the function.

Decision variables

$$x_1, x_2, \cdots x_n \to x = \begin{bmatrix} x_1 \\ x_1 \\ \vdots \\ x_n \end{bmatrix}$$

Where $x$ is the vector of decision variable.

2- Objective function which we want to find its optimum value. In the other words it describes a relation between decision variables and we want to minimize or maximize it.

3- Problem Constraints

$m$ constraints as $=$ or $\leq$ or $\geq$ which we want to be valid at the optimum point.

**Example 5.1:** Consider an objective function of $f(x) = (n - 1)^2$. Minimize $f(x)$ for the problem constraints subject to (S.T.) equal $2 \leq n \leq 3$.

Solution

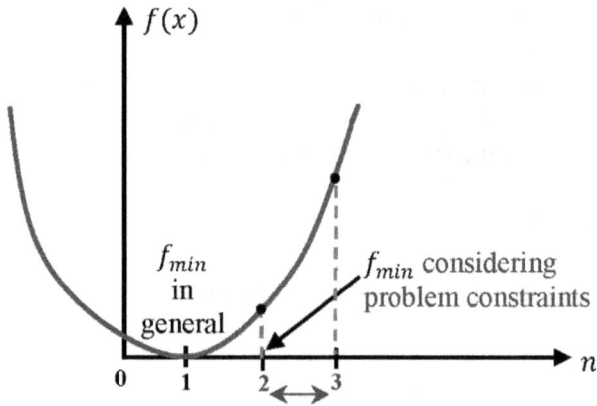

Figure 5.1 Objective function of Example 5.1.

For $n = 2 \rightarrow f^* = (2 - 1)^2 = 1$, min value of $f(x)$ considering constraints.

In general objective function equation (between decision variables) can be a non-linear equation (e.g. 2nd order, 3rd order, sinusoidal,...). In the simplest form of the optimization problem, the objective function describes a linear relationship between decision variables. If both objective function and all of the problem's constraints are linear, it is known as linear programming (LP).

**Example 5.2:** A plant produces three different types of products, as shown in Table 5.1, by applying different operations. Determine how many units of each product should be generated to maximize total plants profit.

Table 5.1 Data of Example 5.2.

| Operation | Operation time to generate one unit of each product (minutes) | | | Daily operation capacity (minutes) |
|---|---|---|---|---|
| | Product 1 $(x_1)$ | Product 2 $(x_2)$ | Product 3 $(x_3)$ | |
| 1 | 1 | 2 | 1 | 430 |
| 2 | 3 | 0 | 2 | 460 |
| 3 | 1 | 4 | 0 | 420 |
| Profit for generation of one unit of each product | 3 | 2 | 5 | |

Solution

The first step of problem formulation is to convert the problem to a mathematical problem by writing a proper mathematical formula that describes the original problem completely).

**1-** Decision variables

$x_1$ = number of generated products 1

$x_2$ = number of generated products 2

$x_3$ = number of generated products 3

**2-** Objective function

Total plants profit is $f(x) = 3x_1 + 2x_2 + 5 x_3$

Where $f(x) = f(x_1, x_2, x_3)$ and $x = \begin{bmatrix} x_1 \\ x_2 \\ x_3 \end{bmatrix}$

**3-** Problem's constraints

For operation 1:  $x_1 + 2x_2 + x_3 \leq 430$

For operation 2:  $3x_1 + 0x_2 + 2x_3 \leq 460$

For operation 3:  $x_1 + 4x_2 + 0x_3 \leq 420$

The other inherent constraints are $x_1, x_2, x_3 \geq 0$ and integer number.

The mathematical form of the problem can be written as:

$$\max f(x) = f(x_1, x_2, x_3) = 3x_1 + 2x_2 + 5x_3$$

S.T.   $x_1 + 2x_2 + x_3 \leq 430$

$$3x_1 + 2x_3 \leq 460$$
$$x_1 + 4x_2 +\leq 420$$
Since $\quad x_1, x_2, x_3 \geq 0$ and integer

There is an analytical solution method for linear programming problems knows as the "simplex method" which can be used to calculate the definitive answer of the problem.

## 5.2 Form Changing of Optimization Problem

### 5.2.1 Conventional Form
An optimization problem in conventional form conditions:
1- All decision variables are non-negative (zero or positive).
2- All constraints are in type less or equal.
3- Objective function is in maximum type.

Any optimization problem can be converted to a conventional form. For example, if the objective function is in minimum type (condition three not met):

$Min\ f(x) \equiv Max\ g(x)$ at $g(x) \triangleq -f(x)$
If condition two is not met:
e.g. $k(x) \geq a \equiv -k \leq -a$
$\quad H(x) = b \equiv H(x) \leq b\ \&\ H(x) \geq b \equiv H(x) \leq b\ \&\ H(x) \leq -b$

Note that the only way to validate both above constraints is $H(x) = b$. If condition one is not met; for example, if we have a variable like $x_1$ without sign constraints, we can replace it with two auxiliary variables like $x_1^+$ and $x_1^-$ which both of them are non-negative:
$x_1 \triangleq x_1^+ - x_1^-$
Where $x_1$ without sign constraint and $x_1^+, x_1^- \geq 0$ with sign constraint.

## Example 5.3
Convert the following problem to conventional form:
Min $\ z_o = 3x_1 - 3x_2 + 7x_3$
S.T. $\ x_1 + x_2 + 3x_3 \leq 40$
$\quad\ x_1 + 2x_2 - 7x_3 \geq 50$
$\quad\quad 5x_1 + 3x_2 = 20$
$x_1^+, x_1^- \geq 0, x_3$ has not to sign constraint

Solution

$x_3^+$ and $x_3^-$ are considered as

$x_3 \triangleq x_3^+ - x_3^-$ and $x_3^+, x_3^- \geq 0$

For objective function $x - 1$

Max $z_o = -3x_1 + 3x_2 - 7(x_3^+ - x_3^-)$

S.T  $x_1 + x_2 + 3(x_3^+ - x_3^-) \leq 40$

$-x_1 - 2x_2 + 7(x_3^+ - x_3^-) \geq -50$

$5x_1 + 3x_2 \leq 20$

$-5x_1 - 3x_2 \leq -20$

$x_1, x_2, x_3^+, x_3^- \geq 0$

## 5.2.2 Standard Form

1- All constraints should be in an equality form (=).
2- The right element of each constraint should be non-negative.
3- All variables should be non-negative.

If the first condition is not met:

e.g., $k(x) \leq a$

$H(x) \geq b$

We consider two auxiliary variables $S_i$ and $S_j$ and add them to the problem as:

$S_i, S_j \geq 0$

$(k(x) \leq a) \equiv (k(x) + S_i = a)$

$(k(x) \geq b) \equiv (H(x) + S_j = b)$

Usually, constraints are numbered and auxiliary variable $S_1, S_2,...$etc., are used for more clarity (variable number selected equal to related constrained number).

If the second condition is not met, the constraint should be multiplied by -1.

If the third condition is not met, the same way as conventional form can be done (e.g. $x_i$ is replaced with $x_i = x_i^+ - x_i^-$).

## Example 5.4

Convert the following problem to standard form:

Min $z_o = 3x_1 - 3x_2 + 7x_3$

S.T.  $x_1 + x_2 + 3x_3 \leq 40$        (1)

$x_1 + 2x_2 - 7x_3 \geq 50$        (2)

$5x_2 - 3x_3 = -20$        (3)

$-6x_1 + 4x_2 \leq -15$        (4)

$x_1, x_2, \geq 0$, $x_3$ has not to sign constraint

Solution

Similar to the previous example, $x_3$ is considered as (based on condition three):

$x_3 \triangleq x_3^+ - x_3^-$ and $x_3^+, x_3^- \geq 0$

For the first constraint, based on condition one, variables $S_1$ is added to the problem:

$$x_1 + x_2 + 3x_3 + S_1 = 40 \quad S_1 \geq 0$$

For 2nd constraint:

$x_1 + 2x_2 - 7x_3 - S_2 = 50$

For 3rd constraint (based on condition two):

$-5x_2 + 3x_3 = 20$

For the 4th constraint (based on conditions one and two):

$-6x_1 + 4x_2 \leq -15$

Thus,

$6x_1 - 4x_2 \leq 15$

And

$6x_1 - 4x_2 - S_4 = 15, S_4 \geq 0$

Find result:

Min $z_o = 3x_1 - 3x_2 + 7(x_3^+ - x_3^-))$

S.T. $x_1 + x_2 + 3(x_3^+ - x_3^-) + S_1 = 40$

$\quad x_1 + 2x_2 - 7(x_3^+ - x_3^-) - S_2 = 50$

$\quad -5x_2 + 3(x_3^+ - x_3^-) = 20$

$\quad 6x_1 - 4x_2 - S_4 = 15$

$x_1, x_2, x_3^+, x_3^-, S_1, S_2, S_4 \geq 0$

## 5.3 Non-linear Optimization Problem Using Lagrange Method

Firstly, consider a problem without any constraints:

$Min\ f(x_1, x_1, \dots, x_n)$

As may be familiar before, at the optimum point, the following conditions should be met.

$$\frac{\partial f}{\partial x_1} = 0$$

$$\frac{\partial f}{\partial x_i} = 0 \Rightarrow \frac{\partial f}{\partial x_2} = 0$$

$$\vdots$$

$$\frac{\partial f}{\partial x_n} = 0$$

- All the above $n$ equations should be met at the optimum point.

140

- These conditions only show the optimum point (max or min can not be detected).

## Example 5.5:
$Min\ f(x_1, x_1) = x_1{}^2 + 4x_2{}^2 - 4x_1{}^2 + 8$

Solution

$$\frac{\partial f}{\partial x_1} = 0 \implies 2x_1 - 4 = 0 \implies x_1 = \frac{4}{2} = 2$$

$$\frac{\partial f}{\partial x_2} = 0 \implies 8x_2 = 0 \implies x_2 = 0$$

Now, the values of $x_1$ and $x_2$ are considered as the optimum point

If the problem including one equation constraint:

$$Min\ f(x_1, x_1, \dots, x_n)$$

S.T. $g(x_1, x_1, \dots, x_n) = 0$ (Note that any equation constraint can be converted to this form)

To solve this problem, the "Lagrange function" is defined as:

$$\mathcal{L} = f - \lambda g$$

Where, $\mathcal{L}$ is the Lagrange function, and $\lambda$ is the Lagrange coefficient.

Really, the problem constraint is added to objective function by a coefficient (an auxiliary variable).

$$d\mathcal{L} \equiv 0$$

$$\implies \frac{\partial f}{\partial x_i} = 0 \qquad\qquad i = 1,2, \dots, n$$

$$\implies \frac{\partial f}{\partial x_i} - \lambda \frac{\partial f}{\partial x_i} = 0 \qquad\qquad i = 1,2, \dots, n$$

Another condition for optimum point is the problems constraint.

$$g(x_1, x_1, \dots, x_n) = 0 \qquad (\equiv \frac{\partial f}{\partial \lambda} = 0)$$

Consequently, at the optimum point

$$\frac{\partial f}{\partial x_i} - \lambda \frac{\partial f}{\partial x_i} = 0 \qquad\qquad i = 1,2, \dots, \qquad (n = \text{equations})$$

$$g(x_1, x_1, \dots, x_n) = 0$$

Totally, $(n + 1)$ conditions

The above equations are called "coordination equations"

## Example 5.6:
$Min\ f(x_1, x_1) = x_1{}^2 + 4x_2{}^2 - 4x_1{}^2 + 8$

S.T. $g = x_1 + 2x_2 - 3 = 0$

Solution

$$d\mathcal{L} = f - \lambda g = x_1{}^2 + 4x_2{}^2 - 4x_1{}^2 + 8 - \lambda(x_1 + 2x_2 - 3)$$

$$\frac{\partial \mathcal{L}}{\partial x_1} = 2x_1 - 4 - \lambda = 0$$

$$\frac{\partial \mathcal{L}}{\partial x_2} = 8x_2 - 2\lambda = 0$$

$$g(x_1, x_2) = x_1 + 2x_2 - 3 = 0$$

By solving this set of equations

$$x_1{}^* = 2.5$$
$$x_2{}^* = 0.25$$
$$\lambda = 1$$

Note that the obtained answer is located on constraint $g = 0$.

## 5.4 Ingredient
- **Objective function**
- **Variables**
- **Constraints**

**Find values of the variables that minimize or maximize the objective function while satisfying the constraints**

## 5.4.1 Different Kinds of Optimization

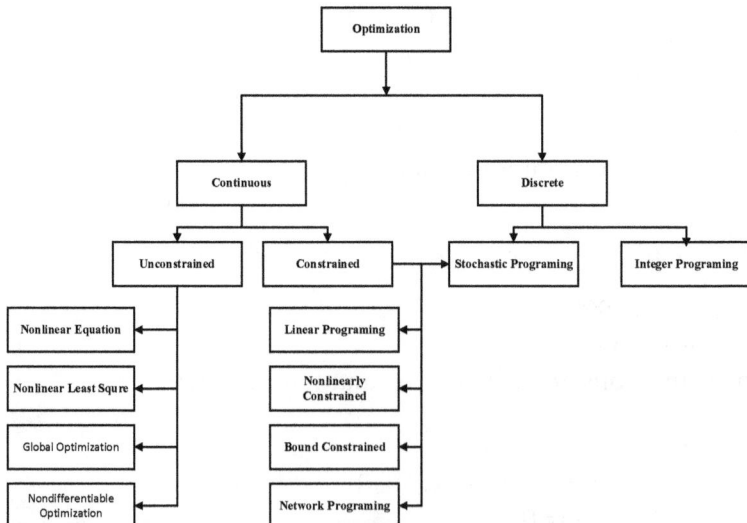

Figure 5.2 Different Kinds of Optimization.

142

### 5.4.2 One-Dimensional Unconstrained Optimization

•     **Root finding** and **optimization** are related. Both involve guessing and searching for a point on a function. Difference is (See Figure 5.3):
- Root finding is searching for zeros of a function
- Optimization is finding the **minimum** or **maximum** of a function of several variables.

In multimodal functions, both local and global optimal can occur as shown in Figure 5.4. We are mostly interested in finding the absolute highest or lowest value of a function.

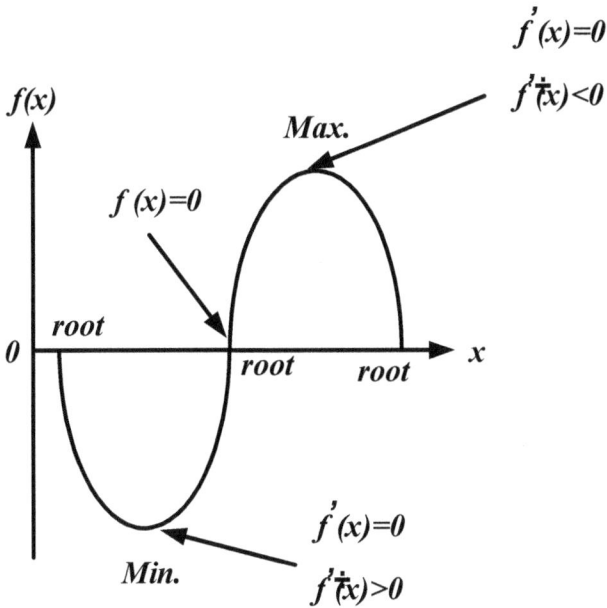

Figure 5.3 roots min. and max. points.

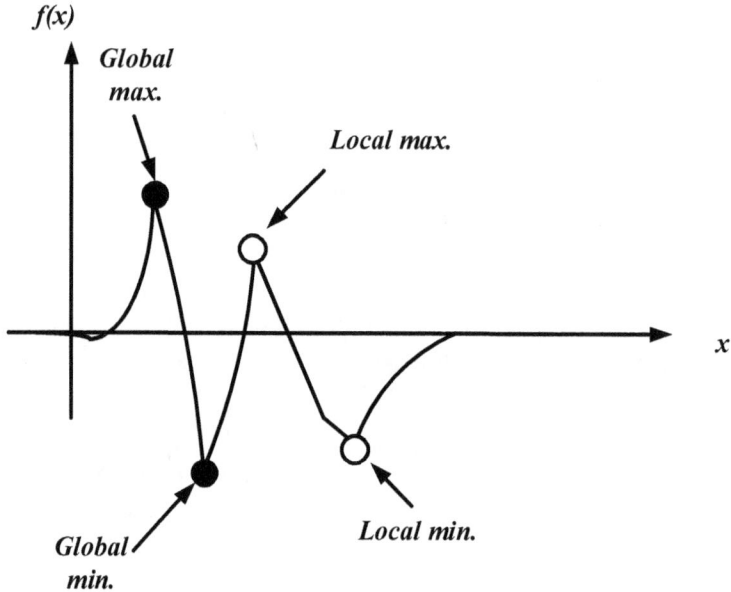

Figure 5.4 local and global optimal points.

A *unimodal* function has a **single maximum** or a **minimum** in the a given interval. For a *unimodal* function :
- First pick two points that will bracket your extremum $[x_l, x_u]$.

Then, pick two more points within this interval to determine whether a **maximum** (See Figure 5.5) has occurred within the *first three* or *last three* points.

$l_0 = l_1 + l_2$ and $\quad \dfrac{l_1}{l_0} = \dfrac{l_2}{l_1}$

$\dfrac{l_1}{l_1 + l_2} = \dfrac{l_2}{l_1} \qquad\qquad R = \dfrac{l_2}{l_1}$

$1 + R = \dfrac{1}{R} \qquad\qquad R^2 + R - 1 = 0$

$R = \dfrac{-1 + \sqrt{1 - 4(-1)}}{2} = \dfrac{\sqrt{5} - 1}{2} = 0.61803$

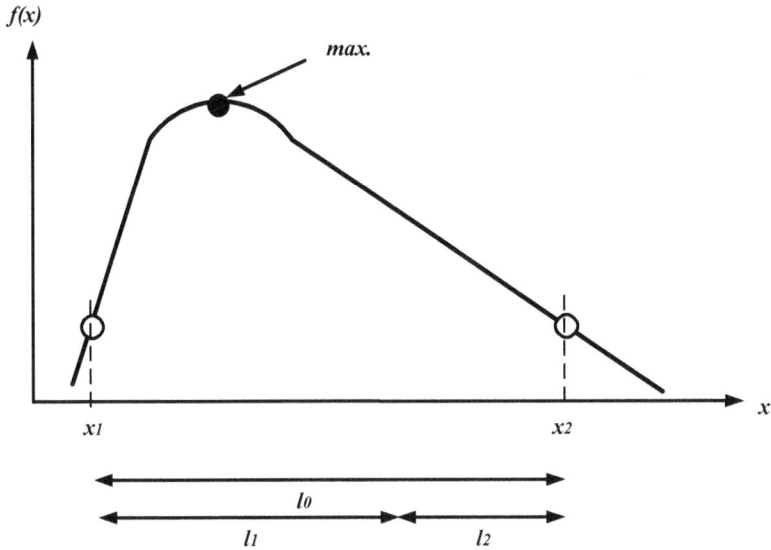

Figure 5.5 Graph shows a max. point.

## 5.5 Golden-Section Search

The Parthenon in Athens, Greece was constructed in the 5th century B.C. It is front dimensions can be fit exactly within a golden rectangle. Figure 5.6 shows the golden rectangle with Parthenon in Athens.

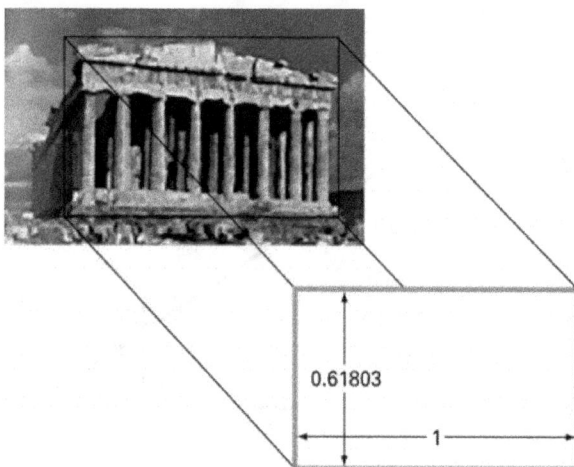

Figure 5.6 The Parthenon in Athens, Greece shows golden rectangle

- Pick two initial guesses, $x_l$ and $x_u$, that bracket one local extremum of $f(x)$ :
- Choose two interior points $x_1$ and $x_2$ according to the **golden ratio** (See Figure 5.7)

$x_1 = x_l + d$
$x_2 = x_u - d$

*Where*

$$d = \frac{\sqrt{5} - 1}{2} \ (x_u - x_l)$$

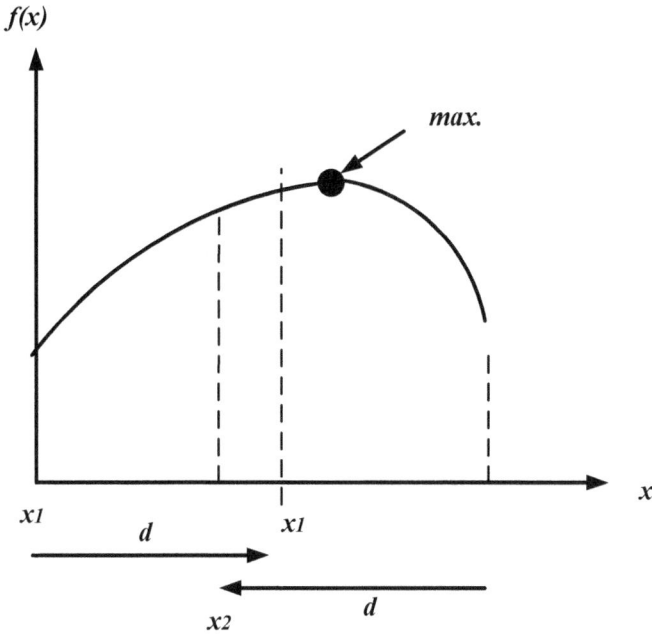

Figure 5.7 Two interior points $x_1$ and $x_2$ according to the **golden ratio.**

**Evaluate the function at $x_1$ and $x_2$:**

- If $f(x_1) > f(x_2)$ then the domain of $x$ to the left of $x_2$ (from $x_1$ to $x_2$) in Figure 5.8 dose not contain the maximum and can be eliminated. Then, $x_2$ becomes the new $x_l$.

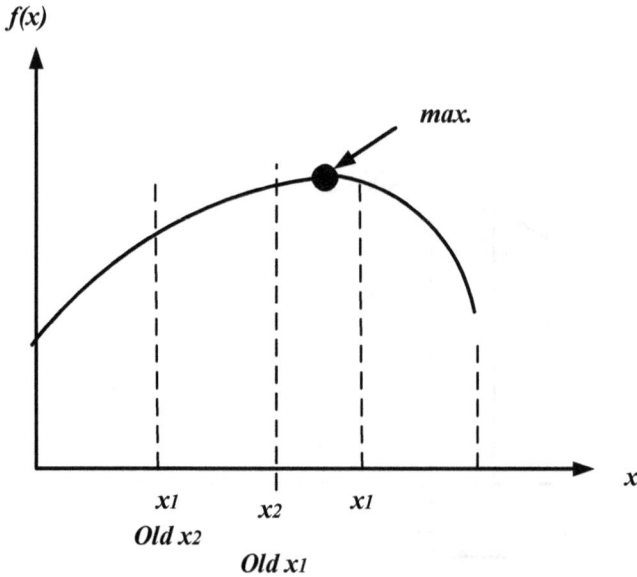

Figure 5.8 The updated of the two interior points $x_1$ and $x_2$ according to the **golden ratio.**

- If $f(x_2) > f(x_1)$, then the domain of x to the right of $x_1$ (from $x_1$ to $x_u$) can be eliminated. In this case, $x_1$ becomes the new $x_u$.
- The benefit of using **golden ratio** is that we do not need to recalculate all the function values in the next iteration.

If $f(x_1) > f(x_2)$ then **New** $x_2 \;\longleftarrow\; x_1$ else **New** $x_1 \longleftarrow x_2$

**Stopping Criteria**

$$|X_u - X_l| < \varepsilon$$

- To accomplish this, we choose relative positions of two points as $\tau$ and $1-\tau$, where $\tau^2 = 1 - \tau$, so $\tau = \frac{(\sqrt{5}-1)}{2} \approx 0.618$ and $1-\tau \approx 0.382$
- Whichever subinterval is retained, its length will be $\tau$ relative to previous interval, and interior point retained will be at position either $\tau$ or $1-\tau$ relative to new interval.
- To continue iteration, we need to compute only one new function value, at complementary point.
- This choice of sample points is called **golden section search**.
- Golden section search (Figure 5.9) is safe but convergence rate is only linear, with constant $C \approx 0.618$.

$\tau = (\sqrt{5} - 1)/2$

$x_1 = a + (1 - \tau)(b - a); f_1 = f_{(x_1)}$

$x_2 = a + \tau(b - a); f_2 = f_{(x_2)}$

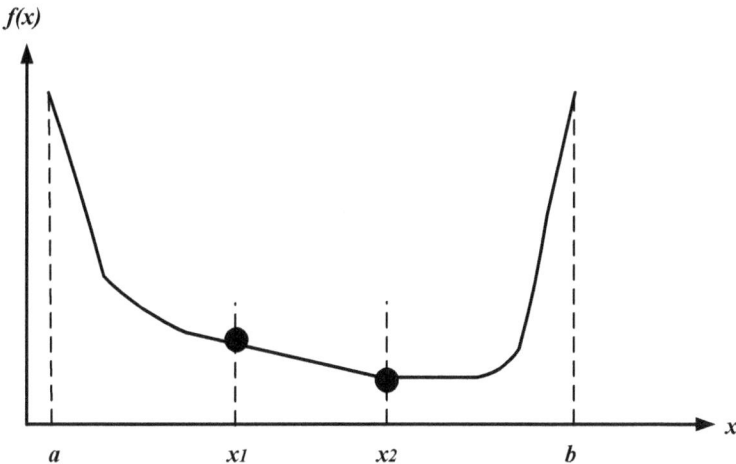

Figure 5.9 Golden section search

**while** $((b - a) > tol.)$ **do**
    **if** $(f_1 > f_2)$ **then**
        $a = x_1$
        $x_1 = x_2$
        $f_1 = f_2$
        $x_2 = a + \tau(b - a)$
        $f_2 = f_{(x_2)}$
    **else**

149

$$b = x_2$$
$$x_2 = x_1$$
$$f_2 = f_1$$
$$x_1 = a+(1 - \tau)(b - a)$$
$$f_1 = f_{(x_1)}$$
  **end**
**end**

## Example 5.1

Use golden section search to minimize
$$f_{(x)} =0.5- x\, exp(-x^2)$$

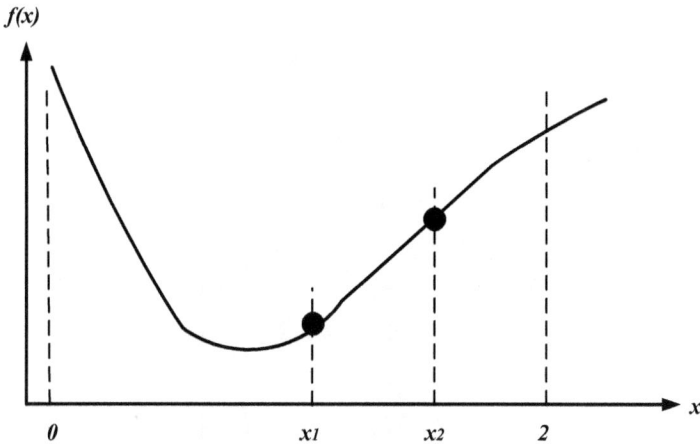

Figure 5.10 Golden section search of Example 5.1.

| $x_1$ | $f_1$ | $x_2$ | $f_2$ |
|-------|-------|-------|-------|
| 0.764 | 0.074 | 1.236 | 0.232 |
| 0.472 | 0.122 | 0.764 | 0.074 |
| 0.764 | 0.074 | 0.944 | 0.113 |
| 0.652 | 0.074 | 0.764 | 0.074 |
| 0.584 | 0.085 | 0.652 | 0.074 |
| 0.652 | 0.074 | 0.695 | 0.071 |
| 0.695 | 0.071 | 0.721 | 0.071 |
| 0.679 | 0.072 | 0.695 | 0.071 |
| 0.695 | 0.071 | 0.705 | 0.071 |
| 0.705 | 0.071 | 0.711 | 0.071 |

## 5.6 Newton's Method
## from Root Finding to Minimization
Second order

approximation of $f(x)$ $f(x) \cong f_{(a)} + \frac{f'(a)}{1!}(x-a) + \frac{f''(a)}{2!}(x-a)^2$

Take derivative (by X)

Minimum is reached
when derivative of
approximation is zero :

$0 = f'(a) + f''(a)(x-a)$

$$x = a - \frac{f'(a)}{f''(a)}$$

- **So.... this is just root finding over the derivative**
  (which makes sense since in local minima, the gradient is zero)
- A similar approach to Newton-Raphson method can be used
to       find       an       optimum       of
$f(x)$ by finding the root of $f'(x)$ (i.e. solving $f'(x)=0$) :

$$X_{i+1} = X_i - \frac{f'(X_i)}{f''(x_i)}$$

- Disadvantage : it may be divergent

**Example 5.2 :**
Use Newton's method to find the maximum of

$f(x) = 2\sin x - x^2/10$ with an initial guess of $x_0 = 2.5$

**Solution :**     $x_{i+1} = x_i - \frac{f'(x_i)}{f''(x_i)} = x_i - \frac{2\cos x_i - x_i/5}{-2\sin x_i - 1/5}$

$x_1 = 2.5 - \frac{2\cos 2.5 - 2.5/5}{-2\sin 2.5 - 1/5} = 0.995$     and     $f(0.995) = 1.578$

Table 5.2 Results

| i | x | f(x) | f'(x) | f''(x) |
|---|------|---------|---------|----------|
| 0 | 2.5 | 0.572 | -2.102 | -1.3969 |
| 1 | 0.995 | 1.578 | 0.8898 | -1.8776 |
| 2 | 1.469 | 1.774 | -0.0905 | -2.1896 |
| 3 | 1.4276 | 1.77573 | -0.0002 | -2.17954 |
| 4 | 1.4275 | 1.77573 | 0.0000 | -2.17952 |

## Problems

5.1  Consider an objective function of $f(x) = (n-2)^2$. Minimize $f(x)$ for the problem constraints subject to (S.T.) equal $2 \le n \le 4$.

5.2  Consider an objective function of $f(x) = (0.5n-2)^2$. Minimize $f(x)$ for the problem constraints subject to (S.T.) equal $2 \le n \le 3$.

5.3 A plant produces three different types of products, as shown in Table 5.3 by applying different operations. Determine how many units of each product should be generated to maximize total plants profit.

Table 5.3 Data of Problem 5.3.

| Operation | Operation time to generate one unit of each product (minutes) | | | Daily operation capacity (minutes) |
|---|---|---|---|---|
| | Product 1 $(x_1)$ | Product 2 $(x_2)$ | Product 3 $(x_3)$ | |
| 1 | 1 | 2 | 1 | 400 |
| 2 | 3 | 0 | 2 | 420 |
| 3 | 2 | 2 | 0 | 440 |
| Profit for generation of one unit of each product | 3 | 2 | 5 | |

5.4 Use golden section search to minimize
$$f_{(x)} = 0.75 - x\, exp(-2x^2)$$

5.5 Find the optimization coordinations
$$Min\ f(x_1, x_1) = x_1^2 + 3x_2^2 - 3x_1^2 + 6$$
$$S.T.\quad g = x_1 + 2x_2 - 2 = 0$$

5.6 Convert the following problem to standard form:

Max $z_0 = 3x_1 - 3x_2 + 7x_3$

S.T.
$$x_1 + x_2 + 2x_3 \le 20 \qquad (1)$$
$$x_1 + 3x_2 - 5x_3 \ge 30 \qquad (2)$$
$$5x_2 - 3x_3 = -10 \qquad (3)$$
$$-5x_1 + 2x_2 \le -25 \qquad (4)$$

$x_1, x_2, \geq 0$, $x_3$ has not to sign constraint

5.7 Convert the following problem to conventional form:

Max $z_o = 2x_1 - 2x_2 + 7x_3$

S.T. $x_1 + x_2 + 2x_3 \leq 20$

$x_1 + 2x_2 - 4x_3 \geq 30$

$5x_1 + 3x_2 = 50$

$x_1^+, x_1^- \geq 0$, $x_3$ has not to sign constraint

# Chapter 6

# Numerical Integration

## 6.1 Rectangular rule:

The integral $\int_a^b f(x).\,dx$ area enclosed by the curve. This area is the sum f area of the rectangles as shown in fig (01), is roughly by the area under the function $f(x)$ from $x = a$; to $x = b$

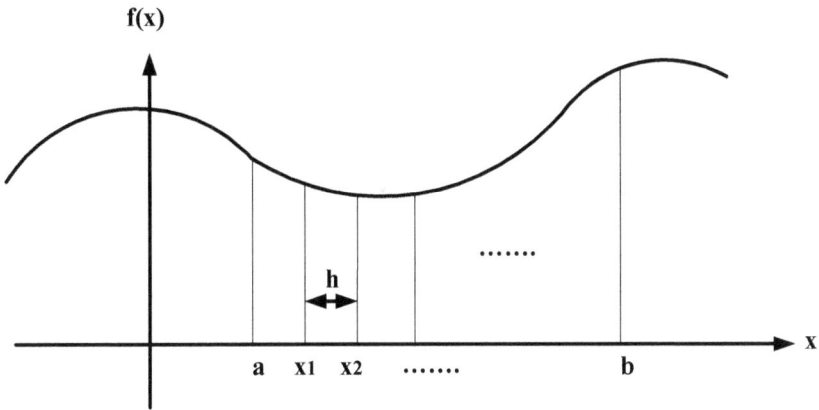

Figure 6.1 Area enclosed by the curve.

$A = \int_a^b f(x).\,dx$

$x_0 = a, x_1 = a + h, x_2 = x_1 + h$

$x_3 = x_2 + h \quad a \, b = a + n.h$

$h = \frac{b-a}{n} \qquad A = \sum_{i=}^{n} h.\,f\left(x_i - \frac{h}{2}\right)$

Example 6.1

Find the approximate value area for $f(x) = x^4$, $a = 0, b = 2$

$$a = 0. b = 2, n = 10. h = 0.2 \ x_0 = \alpha = 0, \ x_1 = 0.2, x_2 =$$

Solution

$0.4x_3 = 0$

$x_4 = 0.8, \ x_5 = 1,$

$$x_6 = 1.2, \ x_7 = 1.4, \ x_8 = 1.6, x_9 = 1.8, \ x_{10} = 2$$

$$A = \sum_{i=1}^{n} h. f(x_i - \frac{h}{2})$$

$$= 0.2[(0.2 - 0.1)^4 + (0.4 - 0.1)^4 + (0.6 - 0.1)^4$$
$$+ (0.8 - 0.1)^4 + (1 - 0.1)^4 + (1.2 - 0.1)^4$$
$$+(1.4 - 0.1)^4 + (1.6 - 0.1)^4 + (1.8 - 0.1)^4 + (2 - 0.1)^4]$$
$$A = 0.2[1.10^{-4} + 8.1 \times 10^{-3} + 6.25 \times 10^{-2} + 240.1 \times 10^{-3}$$
$$+ 656.1 \times 10^{-3}$$
$$+1.4641 + 2.8561 + 5.0625 + 8.3521 + 13.0321]$$
$$A \approx 6.34676 unit$$

While by using the exact method i.e integral we have

$$A = \int_0^{} x^4 \, dx = \frac{2^5}{5} - \frac{0^5}{5} = 6 unit$$

## 6.2    Trapezoidal rule

This rule is used when antiderivative f(x) is not known or f(x) is given in tabular form.

This me the is one f the simplest methods of obtaining an approx. value of the area.

$$h = \frac{b - a}{n}$$

$$A = \frac{h}{2}[y_0 + 2y_1 + 2y_2 + 2y_3 \ldots + 2y_{n-1} + y_1] where$$

$$y_0 = f(x_0) = f(\alpha), y_1 = f(a + h), y_2 = f(x_1 + h), y_3$$
$$= f(x_2 + h), y_4 = f(x_3 + h), y_n = f(b)$$

Example 6.2

Find the approximate value area for $f(x) = x^4$, $a = 0, b = 2$

Solution

$$y_0 = (0)^4 = 0. y_1 = (0.2)^4, y_2 = (0.4)^4, y_3 = (0.6)^4, y_4 = (0.8)^4, y_5$$
$$= (1)^4 y_6 = (1.2)^4$$

$$y_7 = (1.4)^4, y_8 = (1.6)^4, y_9 = (1.8)^4, y_{10} = (2)^4$$

$$A = \frac{0.2}{2}[0 + 3.2 \times 10^{-3} + 0.0512 + 0.2592 + 0.8192 + 2 + 4,1472$$

$$+ 7.6832 + 13.1072 + 20.9952 + 16]$$

$$A \approx 6.50656 unit$$

A better result can be obtained by using a large number of trapezoids

## 6.3 Numerical solution of differential equations

**Example 6.3**

$$y' = \frac{1}{2}(1 + x)y^2, y(0) = 1 \qquad (notlineer)$$

$$Araliticalsol: y = \frac{4}{-2x - x^2}$$

Solution:
Evaluate the given equation by Maclaurin series for the first 4 terms only:

$$y(x) = y_0 + y'_0 + \frac{y''_0}{2}x^2 + \frac{y'''_0}{6}x^3$$

$$y_0 = y(x) at x = 0$$

$$y'_0 = \frac{1}{2}y_0^2 = \frac{1}{2}(y_0^2 = 1)$$

$$y'' = \frac{1}{2}y^2 + (1 + x)y.y' \rightarrow y''' = \frac{1}{2}y.y'$$

$$y'' = \frac{1}{2}y + x)y_0 y'_0 = \frac{1}{2} + (1)(\frac{1}{2}) = 1$$

$$y''' = 2.y.y' + (1 + x)(y^{-2} + yy'')$$

$$y''' = 2.y.y' + (1 + x)(y^{-2} + y_0 y''_0)$$

$$y''' = 2.1.\frac{1}{2} + 1(\frac{1}{4} + 1.1) = 1 + (\frac{1}{4} + 1) = \frac{1}{4}$$

After substitution in the given Maclaurin series, we obtain the follow ire table for different values of X

157

*Table 6.1 Results of Example 6.3*

| X | y | [y] |
|---|---|---|
| 0 | 1 | 1 |
| 0.1 | 1.055375 | 1.055409 |
| 0.2 | 1.123 | 1.123596 |
| 0.3 | 1.205125 | 1.208459 |
| ... | ... | ... |
| ... | ... | ... |

## 6.4    Euler's method

$$y_i' = \frac{changeiny}{changeinx} = \frac{y_{i+1} - y_i}{x_{i+1} - x_i}$$

$(x_{i+1} - x_i)y_i' = y_{i+1} - y_i Euler's formula$
$recursion formula$

$$y_{i+1} = y_i + (x_{i+1} - x_i)y_i'$$
$$= y_i + h. y_i'$$

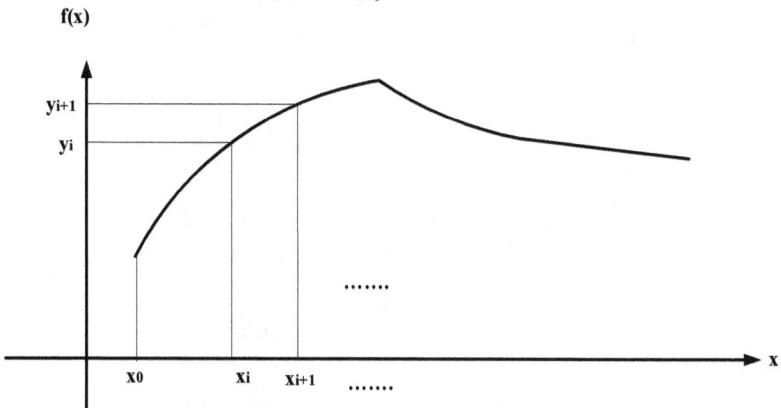

Figure 6.4 Analysis the curve using Euler's method

This is step by step procedure which gives $y_{i+1}$ as  soon a one or more preceding values of $y_i$, $x_{i-1}$, etc...are ... we wish to solve 1st order d.e.

158

$$y' = f(x, y), y(x_0) = y_0$$

Truncation error is of order $h^2$

Where $h$ is the integration interval.

## Example 6.4

Evaluate by Euler's formula $y' = \frac{1}{2}(1 + x)y^2, y(0) = 1$

**Solution:**

$$puti = 0. y_1 = y_0 + h. y_0'$$
$$= 1 + 0.1(\frac{1}{2}(1 + 0)1^2)$$
$$= 1.05$$
$$puti = 1. y_2 = y_1 + h. y_1'$$
$$= 1.05 + 0.2(\frac{1}{2}(1 + 0)1.05^2)$$
$$= 1.1712$$

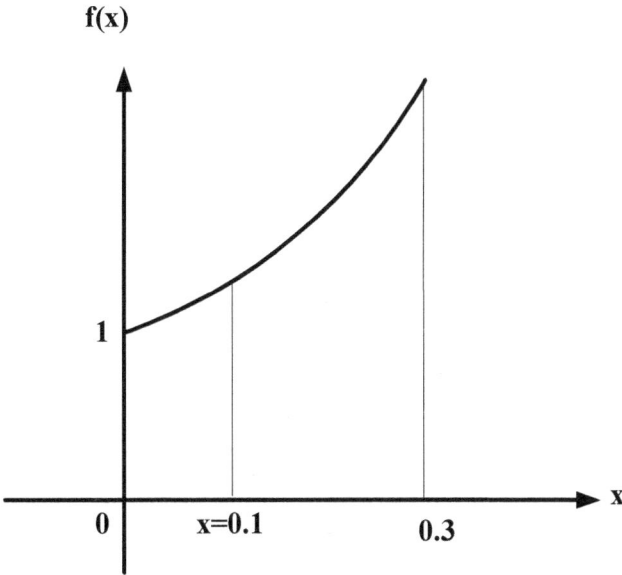

Figure 6.3 Analysis the curve of Example 6.4 using Euler's method.

*Table 6.2 Results of Example 6.4*

| h | i | $x_i$ | $y_i$ | $y_i' = f_i$ | $(x_{i+1} - x_i)f_i$ | [y] |
|---|---|---|---|---|---|---|
| 0.1 | 0 | 0 | 1 | | | 1.0 |
| 0.2 | 1 | 0.1 | 1.03 | 0.606 | | 1.055409 |
| 0.2 | 2 | 3.0 | 1.1712 | | | 1.203459 |
| | | 0.5 | | | | |

$$y_3 = y_2 + h.y_2' = 1.712 + 0.2(\frac{1}{2}(1 + 0.3)(1.712)^2)$$

Example 6.5
Consider the Exponential Decay Example $\dot{x} = -x$ with $x(0) = x_0$
Solution:

This has a solution $x(t) = x_0 e^{-t}$ Since we know the solution we can compare the accuracy of Euler's method for different time steps

*Table 6.3 Results of Example 6.5*

| t | $x^{actual}(t)$ | x(t) Δt=0.1 | x(t) Δt=0.05 |
|---|---|---|---|
| 0 | 10 | 10 | 10 |
| 0.1 | 9.048 | 9 | 9.02 |
| 0.2 | 8.187 | 8.10 | 8.15 |
| 0.3 | 7.408 | 7.29 | 7.35 |
| ... | ... | ... | ... |
| 1.0 | 3.678 | 3.49 | 3.58 |
| ... | ... | ... | ... |
| 2.0 | 1.353 | 1.22 | 1.29 |

Example 6.6

Consider the equations describing the horizontal position of a cart attached to a lossless spring: $\dot{x}_1 = x_2, \dot{x}_2 = -x_1$ Assuming initial conditions of $x_1(0) = 1$ and $x_2(0) = 0$, the analytic solution is $x_1(t) = \cos t$.

Solution:

Starting from the initial conditions at t =0 we next calculate the value of x(t) at time t = 0.25. $x_1(0.25) = x_1(0) + 0.25x_2(0) = 1.0$ $x_2(0.25) = x_2(0) - 0.25x_1(0) = -0.25$ Then we continue on to the next time step, t = 0.50 $x_1(0.50) = x_1(0.25) + 0.25x_2(0.25) = = 1.0 + 0.25 \times (-0.25) = 0.9375$ $x_2(0.50) = x_2(0.25) - 0.25x_1(0.25) = -0.25 - 0.25 \times (1.0) = -0.50$

Table 6.4 Results of Example 6.6

| t | $x_1^{actual}(t)$ | $x_1(t)$  Δt=0.25 |
|---|---|---|
| 0 | 1 | 1 |
| 0.25 | 0.9689 | 1 |
| 0.50 | 0.8776 | 0.9375 |
| 0.75 | 0.7317 | 0.8125 |
| 1.00 | 0.5403 | 0.6289 |
| ... | ... | ... |
| 10.0 | -0.8391 | -3.129 |
| 100.0 | 0.8623 | -151,983 |

## 6.5 Richardson $h^2$ extrapolation:

$$x_{i+1} - x_i = h$$

$y_{i+1}$ is evaluated first by Euler's method with integration. Interval $h$ then evaluated with $\frac{h}{2}$Richardson $h^2$ extrapolation is then employed from an improved value of $y_{i+1}$.

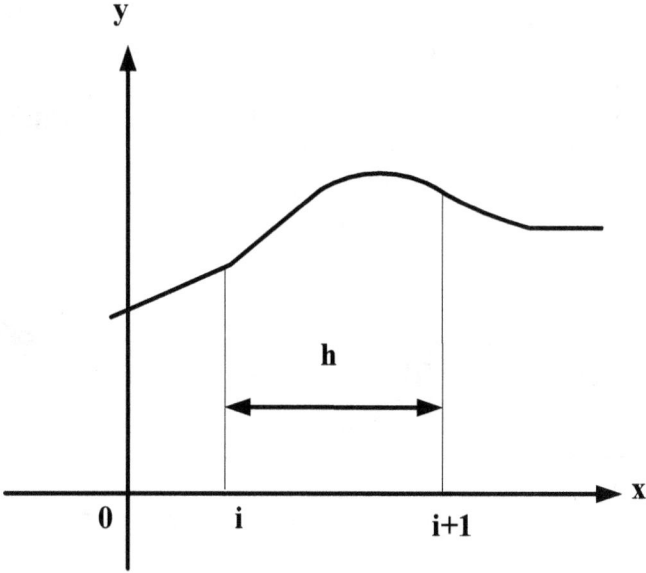

Figure 6.4 Analysis the curve using Richardson $h^2$ extrapolation method.

$$y_{i+1} = y_i + h.f_i$$

$$y_{i+\frac{1}{2}} = y_i + \frac{h}{2}.f_i$$

$$y_{i+1} = y_1 + \frac{1}{2} + \frac{1}{2}.h.f_i + \frac{1}{2}$$

$$y_{i+1} = y_i + \frac{h}{2}(f_i + f_i + \frac{1}{2})$$

↑ improvedvalueof $y_{i+1}$ is

$$y_{i+1} = \frac{4}{3}y_{i+1} - \frac{1}{3}y_{i+1} = y_i + \frac{h}{3}(f_i + 2.f_i + \frac{1}{2})$$

162

**Example 6.7**
Evaluate Example 6.1 by Euler –Richardson method.
**Solution**:

$$y_{i+\frac{1}{2}} = y_i + \frac{h}{2}.f_i = 1 + \frac{0.05}{2} = 1.025$$

*Table 6.5 Results of Example 6.7*

| $x_i$ | $x_{i+\frac{1}{2}}$ | $y_i$ | $f_i$ | $y_{i+\frac{1}{2}}$ | $f_{i+\frac{1}{2}}$ | $\frac{h}{3}(f_i + 2.f_i + \frac{1}{2})$ | $[y]$ |
|---|---|---|---|---|---|---|---|
| 0 | | 1 | 0.5 | | | | 1 |
| | 0.05 | | | 1;025 | 0.552 | | |
| 0.1 | | 1.0535 | 0.6104 | | | | 1.0554 |
| | 0.2 | | | 1.1145 | | | |
| 0.3 | | 1.1942 | | | | | 1.208459 |

$$y_{i+1} = y_{i+1} + a.h^2 f_{i+\frac{1}{2}} = y'_{i+\frac{1}{2}}$$

$$y_{i+1} = y_{i+1} + a.\frac{h^2}{4}f_{i+\frac{1}{2}} = y'_{i+\frac{1}{2}} = \frac{1}{2}(1 + y_{i+1})(y_{i+\frac{1}{2}})^2$$

$$f_{i+\frac{1}{2}} = \frac{1}{2}(1 + 0.05)(1.025)^2 = 0.552$$

Elimination (a) to obtain the re…. formula.

## 6.6 Rung – Kutta method

High order d.e's could be broken down to a system eqs of first order.

$$y(x) = y_0 + \frac{y'_0}{2} + \frac{y''_\circ}{3!} + \dots$$

$$y'_j = f_j(x, y) \qquad j = 1.2, \dots\dots\dots\dots\dots, n$$

## 2nd order Rung-Kutta method
$$\Delta'y_i = h.f_i = h.f(x_i, y_i)\Delta'y_i: \ inverse \ mntf, g$$

$$\overline{\Delta}y_i = h.f(\overline{x}, \overline{y})$$

$$= h.f\left(x_i + \frac{h}{2}, y_i + \frac{1}{2}\Delta'y_i\right) \overline{\Delta}y_i : total\ inverse\ mnt\ truncation\ error$$

is of order $h^3$

$$y_{i+1} = y_i + \overline{\Delta}y_i$$

**3rd order R.K method**

$$\Delta'y_i = h.f(x_i, y_i)$$

$$\Delta''y_i = h.f\left(x_i + \frac{h}{2}, y_i + \frac{1}{2}.\Delta'y_i\right) truncation\ error$$

is of order $h^4$

$$\Delta'''y_i = h.f(x_i + h, y_i + 2.\Delta''y_i - \Delta'y_i)$$

$$\overline{\Delta}y_i = \frac{1}{6}(\Delta'y_i + 4\Delta''y_i + \Delta'''y_i)$$

$$y_{i+1} = y_i + \overline{\Delta}y_i$$

**4th order R.K method**

$$\Delta'y_i = h.f(x_i, y_i)$$

$$\Delta''y_i = h.f(x_i + \frac{h}{2}, y_i + \frac{1}{2}\Delta'y_i)$$

$$\Delta'''y_i = h.f\left(x_i + \frac{h}{2}, y_i + \frac{1}{2}.\Delta''y_i\right) truncation\ error$$

is of order $h^5$

$$\Delta''''y_i = h.f(x_i + h, y_i + \Delta'''y_i)$$

$$\overline{\Delta}y_i = \frac{1}{6}(\Delta'y_i + 2.\Delta''y_i + 2.\Delta'''y_i + \Delta''''y_i)$$

$$y_{i+1} = y_i + \overline{\Delta}y_i$$

**Example 6.8**

solve Example 6.1 by 2nd order R.K method

$$x = 0.1$$

$$y' = \frac{1}{2}(1 + x)y^2, y(0) = 1$$

**Solution:**

$$f = y' = \frac{1}{2}(1 + x)y^2$$

164

$$= \frac{1}{2}(1 + 0.05)(1.025)h.f(\overline{x}, \overline{y})$$

Table 6.6 Results of Example 6.8

| [y] | $x_i$ | $y_i$ | $f(x_i, y_i)$ | $\frac{1}{2}\Delta'y_i$ | $x_i + \frac{h}{2}$ | $y_i + \frac{1}{2}\Delta'y_i$ | $f(\overline{x}, \overline{y})$ | $\overline{\Delta}y_i$ |
|---|---|---|---|---|---|---|---|---|
| 1.0 | 0.0 | 1.0 | 0.5 | 0.025 | 0.05 | 1.025 | 0.5516 | 0.0552 |
| 1.0554 | 0.1 | 1.0552 | 0.6124 | 0.0612 | 0.02 | 1.116 21 | 0.7478 | 0.1496 |
| 1.2085 | 0.3 | 1.2048 | | | | | | |

## Example 6.9
Solve Example 6.1 by 3rd O.R.K method for $x = 0 - 1$ only.
**Solution:**
Table 6.7 Results of Example 6.9

| [y] | $x$ | $y_i$ | EulerRicgar | R.k 2$^{nd}$ orde | R.k 3$^{nd}$ orde | R.k 4$^{nd}$ orde | Euler |
|---|---|---|---|---|---|---|---|
| 1.05549 | 0.1 | 1.055375 | 1.0535 | 1.055 16 | 1.055 411 | 1.055 409 | 1.05 5 |

$$\Delta'y_i = h.f_i = 0.1 \times 0.5 = 0.05$$

$$\Delta''y_i = 0.1\left[\frac{1}{2}(1 + 0.05)(1.025)^2\right] = 0.055158$$

$$\Delta'''y_i = h.f(x_i + h, y_i + 2.\Delta''y_i - \Delta'y_i)$$

$$\Delta''''y_i = 0.1\left[\frac{1}{2}(1 + 0.1)(1.060316)^2\right]$$

$$\overline{\Delta}y_i = \frac{1}{6}[0.05 + 4(0.055158) + 0.061835]$$
$$= 1.055411$$

**Example 6.10**
**Evaluate**

$$y' - x(1-x)y = 0$$

$$y(0) = 10. \, h = 0.2$$

a-      By Euler's method
b-      2nd order R.K. method for $x = 0.2; 0.4; 0.5$

**Solution:**

$$y_{i+1} = y_i + h.f_i$$
$$i = 0; x = 0; \quad y = 10$$
$$y_1 = 10 + 0.2 \ x0 = 10$$
$$i = 1. \ x = 0.2, \quad y = 10$$
$$y_2 = y_1 + h.f_1$$
$$= 10 + 0.2(1.6) = 10.32$$

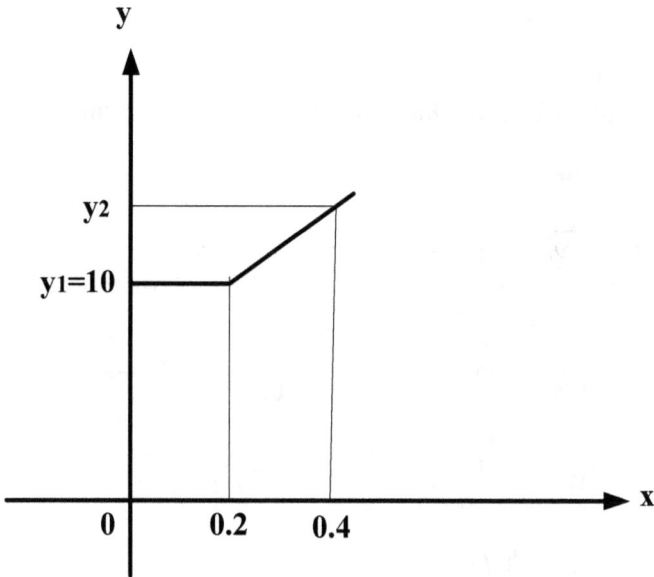

Figure 6.5 Analysis the curve of Example 6.8

$$i = 2, \quad x = 0.4, \quad y = 10.32$$

$$y_3 = y_2 + h.f_2 = 10.32 + 0.2. \, f_2 f_2 = 0.4.10.32(1 - 0.4)$$
$$= 0.4(10.32)(0.6)$$

$$\Delta' y_i = h.f_i = h.f(x_i, y_i)$$

$\Delta' y_0 = 0.2. x_0$

$$\overline{\Delta y_0} = h. f(\overline{x}, \overline{y}) = h. f\left(x_0 + \frac{h}{2}, y_0 + \frac{1}{2}. \Delta' y_0\right) = 0.2[0.1; 10]$$
$$= 0.18$$

$y_1 = y_0 + \overline{\Delta y_0} = 10.18$

$$\Delta' y_1 = \frac{h}{0.2}(0.2(10.18(1 - 0.2)) = 0.32576$$

$\overline{\Delta y_1} = 0.2(0.3(10.18 + 16288)(1 - 0.3))$

$= 0.2(0.3)(10.3428) = 0.4344$

$y_2 = y_1 + \overline{\Delta y_1} = 10.18 + 0.4344$

$= 10.6144$

## Example 6.11

Consider the same example from before the position of a cart attached to a lossless spring. Again, with initial conditions of $x_1(0)$ =1 and $x_2(0) = 0$, the analytic solution is $(t) = cos(t)$

$$\dot{x}_1 = x_2$$
$$\dot{x}_2 = -x_1$$

With $\Delta t = 0.25$ $at$ $t = 0$

## Solution

$$k_1 = (0.25) \times \begin{bmatrix} 0 \\ -1 \end{bmatrix} = \begin{bmatrix} 0 \\ -0.25 \end{bmatrix}$$
$$x(0) + k1 = \begin{bmatrix} 1 \\ 0 \end{bmatrix} + \begin{bmatrix} 0 \\ -0.25 \end{bmatrix} = \begin{bmatrix} 1 \\ -0.25 \end{bmatrix}$$
$$k_2 = (0.25) \times f(x(0) + k_1) = \begin{bmatrix} -0.0625 \\ -0.25 \end{bmatrix}$$
$$x(0.25) = \begin{bmatrix} 1 \\ 0 \end{bmatrix} + \frac{1}{2}(k_1 + k_2) = \begin{bmatrix} 0.96875 \\ -0.25 \end{bmatrix}$$

Table 6.8 compares the numeric and exact solutions for $x_1(t)$ using the RK2 algorithm

*Table 6.8 Results of Example 6.11*

| time | actual $x_1(t)$ | $x_1(t)$ with RK2 $\Delta t=0.25$ |
|---|---|---|
| 0 | 1 | 1 |
| 0.25 | 0.9689 | 0.969 |
| 0.50 | 0.8776 | 0.876 |
| 0.75 | 0.7317 | 0.728 |
| 1.00 | 0.5403 | 0.533 |
| 10.0 | -0.8391 | -0.795 |
| 100.0 | 0.8623 | 1.072 |

## Problems

6.1 Find the approximate value area for $f(x) = x^4$ , $a = 0, b = 2$
$a = 0. b = 2, n = 10. h = 0.2 \; x_0 = \alpha = 0, \quad x_1 = 0.2, x_2 = 0.4, x_3 = 0$

6.2 Evaluate by Euler's formula $y' = \frac{1}{2}(1 + x)y^2, y(0) = 1.5$

6.3 Evaluate by Euler's formula $y' = \frac{1}{2}(2 + x)y^2, y(0) = 2$

6.4 Evaluate by Euler's formula $y' = (1 + 2x)y^2, y(0) = 1$

6.5 Evaluate using Euler –Richardson method.
$f(x) = x^4$ , $a = 0, b = 2$
$a = 0. b = 2, n = 10. h = 0.2 \; x_0 = \alpha = 0, \quad x_1 = 0.2, x_2 = 0.4, x_3 = 0$

6.6 solve by 2nd order R.K method
$$x = 0.1$$
$$y' = \frac{1}{2}(2 + x)y^2, y(0) = 2$$

# Chapter 7

# The Laplace Transform

The Laplace transform is introduced in this chapter as a highly strong alternative tool for studying systems. The Laplace transform is named after the French astronomer and mathematician Pierre Simon Laplace (1749-1827). In contrast to the time-domain models discussed in the preceding chapter, the Laplace transform is a frequency-domain representation that simplifies linear system analysis and design.

The Laplace transform is a well-known technique in the analysis of continuous-time, linear systems. It is significant for a number of reasons. For starters, it is applicable to a broader range of inputs. It is possible to find the Laplace transform of an unbounded signal. Second, the Laplace transform is powerful for supplying us with the whole response, that is, the steady-state plus transient or homogeneous, in a single operation, and third, it automatically incorporates the beginning circumstances in the system analysis. It enables us to transform ordinary differential equations into algebraic equations that are easier to manage and solve. It reduces convolution to basic multiplication. Fourth, we may use the Laplace transform to build a continuous-time LTI system's transfer function representation.

The chapter begins with the definition of the Laplace transform and uses that definition to derive the transform of some basic, important functions. We will consider some properties of Laplace transform which are helpful in obtaining the Laplace transform of other functions. We then consider the inverse Laplace transform. We apply all these to solving integrodifferential equations and system analysis, especially in circuit and control systems. We finally demonstrate how MATLAB can be used to do most of what we cover in this chapter.

## 7.1 Definition

There are two types of Laplace transforms.
1) The two-sided (or bilateral) Laplace transform

The two-sided Laplace transform allows time functions to be non-zero for negative time.

$$\mathcal{L}[x(t)] = X(s)$$
$$= \int_{-\infty}^{\infty} x(t)e^{-st}\, dt \qquad (7.1)$$

where s is the complex frequency given by
$$s = \sigma + j\omega. \qquad (7.2)$$
where $\sigma$ (in nepers/s) and $\omega$ (in radians/s) are the real and imaginary parts of s respectively. The bilateral Laplace transform advantage is that it can handle both causal and non-causal signals over -∞ to ∞.

2) The one-sided (or unilateral) form
   A one-sided function is zero for negative time; that is, - t < 0 , where - 0 denotes a time just before 0. It is found from Equation (7.1) by setting the lower limit of the integral equal to zero.

$$\mathcal{L}[x(t)] = X(s) = \int_{0^-}^{\infty} x(t)e^{-st}\, dt \qquad (7.3)$$

The one-sided Laplace transform is most commonly used transform than the two-sided Laplace transform because here we encounter only positive-time signals in practice and it is defined for positive-time signals.

According to Equation (7.3), the t-domain signal x(t) is changed into the s-domain function X(s).

The **Laplace transform** of a signal x(t) is the integration of the product of x(t) and $e^{-st}$ over the interval from 0 to +∞.

In inverse Laplace transform the s-domain X(s) is changed back to t-domain x(t) and is given as

170

$$\mathcal{L}^{-1}[X(s)] = x(t)$$

$$= \frac{1}{2\pi j} \int_{\sigma_1-j\infty}^{\sigma_1+j\infty} X(s)e^{st}ds \tag{7.4}$$

where the integration is performed along a straight line ($\sigma_1 + j\omega$, $-\infty < \omega < \infty$), as shown in **Figure7.1**.

The Laplace transform pair x(t) and X(s) are represented as

$$x(t) \iff X(s) \tag{7.5}$$

because the transform goes both ways and that there is one-to-one correspondence between x(t) and X(s).

A signal x(t) is Laplace transformable if the integral in Equation (7.3) exists; i.e. in order for x(t) to have a Laplace transform, the integral in Equation (7.3) must converge. The integral converges when

$$\left| \int_{0^-}^{\infty} x(t)e^{-(\sigma+j\omega)t}dt \right| < \infty$$

$$\tag{7.6}$$

This can be simplified as

$$\left| \int_{0^-}^{\infty} x(t)e^{-(\sigma+j\omega)t}dt \right| \leq \int_{0^-}^{\infty} \left| x(t)e^{-(\sigma+j\omega)t}dt \right|$$

$$\leq \int_{0^-}^{\infty} |x(t)| \left| e^{-(\sigma+j\omega)t}dt \right| < \infty$$

$$\tag{7.7}$$

Since $|e^{j\omega t}| = 1$ for any value of t,

$$\int_{0^-}^{\infty} |x(t)|e^{-\sigma t}dt < \infty \tag{7.8}$$

for some real value of $\sigma = \sigma_c$. The inequality in Equation (7.8) states that the Laplace transform exists. Hence, the region of convergence (ROC) for Laplace transform is Re(s) = $\sigma > \sigma_c$, as shown in **Figure7.1**, the region of convergence means, the range of s for which the Laplace transform converges. We cannot define X(s) outside the region of convergence. The portion of the s plane in which the integral converges is found by the value of $\sigma$. All signals that we come across within the system analysis fulfills the convergence criterion in Equation 7.8 and have their corresponding Laplace transforms.

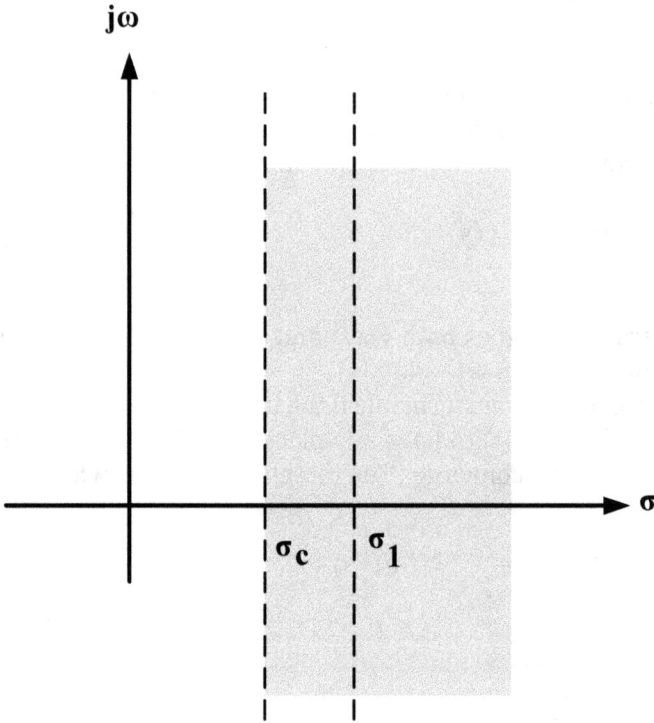

Figure 7.1 Region of convergence for the Laplace transform.

**Example 7.1**  Find the Laplace transform of the following functions and establish the region of convergence (ROC) for each case.
(a) $e^{-5t} u(t)$,
(b) $\delta(t)$.

**Solution**
    (a) For the exponential function, shown in **Figure** 7.2(a), the Laplace transform is

$$L[e^{-5t}u(t)] = \int_{0^-}^{\infty} e^{-5t}e^{-st}\,dt = -\frac{1}{s+5}e^{-(s+5)t}\Big|_{0^-}^{\infty}$$

$$= \frac{1}{s+5} \tag{7.8. a}$$

Notice that $\lim\limits_{t\to\infty} e^{-(s+5)t} = 0$ only if $Re(s+5) > 0$ or $Re(s) > -5$. Alternatively, the ROC is obtained from

$$|e^{-(s+5)t}| = |e^{-(\sigma+5)t}||e^{-j\omega t}| = |e^{-(\sigma+5)t}| < \infty$$

which is valid when $\sigma + 5 > 0$ or $\sigma > -5$. The result in Equation (7.8.b) is so important that it deserves boxing.
In general

$$L[e^{-at}u(t)] = \frac{1}{s+a} \tag{7.8. b}$$

A special case of this is when a=0. $e^{-at}u(t)$ becomes u(t) and we have

$$L[u(t)] = \frac{1}{s}$$

$$\tag{7.8.c}$$

(b) For the unit impulse function, shown in **Figure7.2(b)**,

$$L[\delta(t)] = \int_{0^-}^{\infty} \delta(t)e^{-st}\,dt = e^{-0} = 1 \quad \text{for all s} \tag{7.8.d}$$

The sifting property has been applied in Equation (7.8.d).In this case, the transform does not depend on s and hence the region of convergence is the entire s plane.

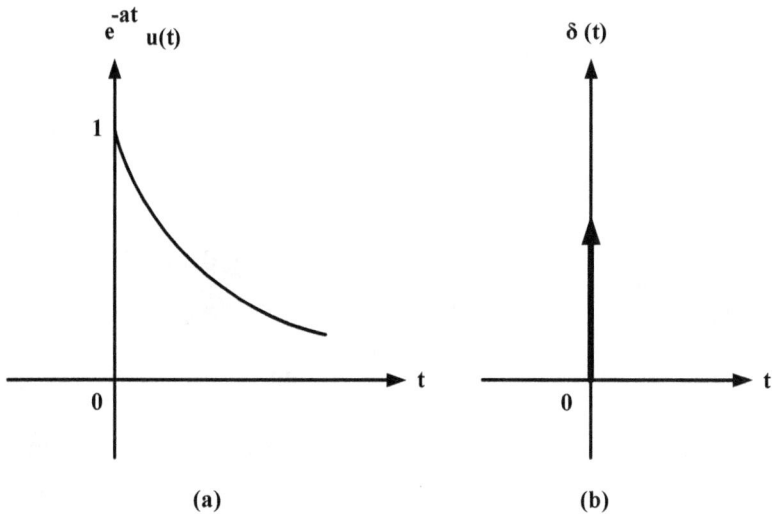

(a)                                    (b)

**Figure** 7.2 Functions of Example 7.1 (a) Exponential function, (b) unit impulse function.

**Exercise** Determine the Laplace transform of these functions and specify the region of convergence for each:

(a) $h(t) = 0.5\, t^2\, u(t)$     (b) $e^{-j\omega t}u(t)$

**Example7.2**  Find the Laplace transform of $x(t) = sin\, \omega t\, u(t)$.

**Solution**

The Laplace transform of the cosine function is

$$X(s) = L[sin\, \omega\, t] = \int_0^\infty (sin\, \omega\, t)e^{-st}dt$$

$$= \int_0^\infty \left(\frac{e^{j\omega t} - e^{-j\omega t}}{2j}\right)e^{-st}dt$$

$$= \frac{1}{2j}\int_0^\infty \left(e^{-(s-j\omega)t} - e^{-(s+j\omega)t}\right)dt$$

$$= \frac{1}{2j}\left(\frac{1}{s-j\omega} - \frac{1}{s+j\omega}\right) = \frac{\omega}{s^2 + \omega^2}$$

## 7.2 Properties of the Laplace Transform

The properties of the Laplace transform are summarized below which are useful to obtain the transform of many signals. Some of the properties are linearity, Scaling, Time shifting, Frequency shifting, Time Differentiation, Time Convolution, Time Integration, Frequency Differentiation , Time Periodicity, Modulation and Initial and Final Values.

### 7.2.1. Linearity

Let the Laplace transform of two signals $x_1(t)$ and $x_2(t)$ are $X_1(s)$ and $X_2(s)$ respectively. Then the linearity property is defined as the Laplace transform of the sum, of two functions of time is equal to the sum of the transforms of each function and is given as

$$\mathcal{L}[a_1 x_1(t) + a_2 x_2(t)] = a_1 X_1(s) + a_2 X_2(s) \tag{7.9}$$

where $a_1$ and $a_2$ are constants.

The linearity property is explained using the following example.

$$\mathcal{L}[a_1 x_1(t) + a_2 x_2(t)]$$
$$= \int_0^\infty [a_1 x_1(t) + a_2 x_2(t)] e^{-at} dt$$

$$= a_1 \int_0^\infty x_1(t) e^{-at} dt + a_2 \int_0^\infty x_2(t) e^{-at} dt \tag{7.10}$$

$$= a_1 X_1(s) + a_2 X_2(s)$$

175

**Example**

By using linearity property ,find $X(s)$
$$x(t) = 3e^{4t} + 5e^{-6t}$$

Solution:
$$X(s) = L[3e^{4t} + 5e^{-6t}] = L[3e^{4t}] + L[5e^{-6t}]$$
$$X(s) = \frac{3}{s-4} + \frac{5}{s+6}$$

## 7.2.2. Scaling

Let X(s) be the Laplace transform of the signal x(t). Let $a$ be any positive real constant, then Laplace transform of x($at$) is given as

$$\mathcal{L}[x(at)] = \int_0^\infty x(at)\, e^{-st}\, dt \tag{7.11}$$

Let λ=$at$, dλ = $a$ dt, we obtain

$$\mathcal{L}[x(at)]$$
$$= \int_0^\infty x(\lambda)\, e^{-\lambda(\frac{s}{a})} \frac{d\lambda}{a}$$
$$= \frac{1}{a} \int_0^\infty x(\lambda)\, e^{-\lambda(\frac{s}{a})}\, d\lambda \tag{7.12}$$

Thus, the scaling property is obtained as

$$\mathcal{L}[x(at)] = \frac{1}{a} X\left(\frac{s}{a}\right), \quad a > 0 \tag{7.13}$$

The unilateral Laplace transform is valid only for causal signals and not valid for non-causal signal (when $a$ is negative i.e., $a < 0$).

**Example:**
Consider
$$L[r(t)] = \frac{1}{s^2} \tag{7.14}$$

By applying the scaling property,

$$L[r(2t)] = \frac{1}{2} \frac{1}{(s/2)^2} = \frac{2}{s^2} \qquad (7.15)$$

### 7.2.3 Time shifting

Often we can see that there is a signal delay in control systems due to transmission delays or other effects. Let X(s) be the Laplace transform of signal x(t). Then time shifting is given as

$$\mathcal{L}[x(t-a)u(t-a)] = e^{-as} X(s) \qquad (7.16)$$

If a function is delayed in time by $a$, then the Laplace transform of the function is multiplied by $e^{-as}$.

Below example describes the time shifting property

$$\mathcal{L}[x(t-a)u(t-a)]$$
$$= \int_0^\infty x(t-a)u(t-a) e^{-st} dt, \quad a$$
$$\geq 0 \qquad (7.17)$$

If $u(t-a) = 0$ for $t < a$ and $u(t-a) = 1$ for $t > a$.

Then,

$$\mathcal{L}[x(t-a)u(t-a)]$$
$$= \int_0^a 0 e^{-st} dt$$
$$+ \int_a^\infty x(t$$
$$- a)(1)e^{-st} dt \qquad (7.18)$$

By substituting $\lambda = t - a$,    $d\lambda = dt$ and $t = \lambda + a$. As    $t \to a$,    $\lambda \to 0$ and as $t \to \infty$,    $\lambda \to \infty$.

Thus,

177

$$\mathcal{L}[x(t-a)u(t-a)] = \int_0^\infty x(\lambda)\, e^{-s(\lambda+a)}d\lambda$$

$$= e^{-as} \int_0^\infty x(\lambda)\, e^{-s\lambda}d\lambda$$

$$= e^{-as}X(s)$$

$$(7.19)$$

### Example

Consider $g(t) = u(t-4) - u(t-6)$.
We know that

$$\mathcal{L}[u(t)] = \frac{1}{s}$$

Using linearity property and the time shifting property,

$$G(s) = L[u(t-4) - u(t-6)] = \frac{1}{s}e^{-4s} - \frac{1}{s}e^{-6s}$$

$$= \frac{1}{s}(e^{-4s} - e^{-6s})$$

$$(7.20)$$

### 7.2.4. Frequency shifting

Let X(s) be the Laplace transform of the signal x(t). Then the frequency shifting is given as

$$\mathcal{L}[e^{-at}x(t)u(t)]$$

$$= \int_0^\infty e^{-at}x(t)e^{-st}dt$$

$$= \int_0^\infty x(t)e^{-(s+a)t}\ dt = X(s+a)$$

$$\mathcal{L}[e^{-at}x(t)u(t)] = X(s+a)$$

$$(7.21)$$

The Laplace transform turns multiplication by $e^{-at}$ in the time domain into s+a in the transform X(s).

**Example**

Consider

$$\cos \omega t \cdot u(t) \quad \Leftrightarrow \quad \frac{s}{s^2 + \omega^2}$$

(7.22)

By applying the frequency property and replace every s by s+a , then Laplace transform of the cosine function is given by

$$L[e^{-at} \cos \omega t u(t)] = \frac{s + a}{(s + a)^2 + \omega^2}$$

(7.23)

## 7.2.5 Time Differentiation

Let X(s) be the Laplace transform of the signal x(t). Time differentiation property is important for solving differential equations. The derivative of x(t) and its Laplace transform is given as

$$L\left[\frac{dx}{dt} u(t)\right] = \int_{0^-}^{\infty} \frac{dx}{dt} e^{-st} dt$$

(7.24)

By derivative by parts, $U = e^{-st}, dU = -se^{-st} dt$ and $dV = \left(\frac{dx}{dt}\right) dt = dx, V = x(t),$then

$$L\left[\frac{dx}{dt} u(t)\right] = x(t)e^{-st}|_{0^-}^{\infty} - \int_{0^-}^{\infty} x(t)[-se^{-st}]dt$$

$$= 0 - x(0^-) + s \int_{0^-}^{\infty} x(t)e^{-st}dt$$

$$= sX(s) - x(0^-)$$

179

$$\mathcal{L}[x'(t)] = sX(s) - x(0^-) \qquad (7.25)$$

By differentiating the above equation, we get

$$\mathcal{L}\left[\frac{d^2x}{dt^2}\right] = s\mathcal{L}[x'(t)] - x'(0^-)$$

$$= s[sX(s) - x(0^-)] - x'(0^-)$$

$$= s^2X(s) - sx(0^-) - x'(0^-)$$

$$\mathcal{L}[x''(t)] = s^2X(s) - sx(0^-) - x'(0^-)$$

$$(7.26)$$

In the same way, for the nth derivative, we get

$$\mathcal{L}\left[\frac{d^nx}{dt^n}\right] = s^nX(s) - s^{n-1}x(0^-) - s^{n-2}x'(0^{-1}) - L$$
$$- s^0 x^{(n-1)}(0^-)$$

$$(7.27)$$

**Example:**

Consider $x(t) = \cos \omega t \, u(t)$

Then at $t = 0$, we have $x(0) = 1$.
By applying time differentiation, the Laplace transform of x(t) is

$$L\left[\frac{d}{dt}\cos \omega t u(t)\right] = s\left(\frac{s}{s^2+\omega^2}\right) - \cos(0^-)u(0^-)$$

$$= \frac{s^2}{s^2+\omega^2} \qquad (7.28)$$

**7.2.6 Time Convolution**

Let X(s) and H(s) be the Laplace transform of the signal x(t) and h(t). Then the convolution property is given as

180

$$\mathcal{L}[x(t) * h(t)] = X(s)H(s) \tag{7.29}$$

In time domain the convolution property,

$$x(t) * h(t) = \int_0^\infty x(t - \tau)h(\tau)d\tau$$

Considering the Laplace transform on both sides

$$\mathcal{L}[x(t) * h(t)] = \int_0^\infty \left[\int_0^\infty h(\tau)x(t - \tau)d\tau\right]e^{-st}dt$$

$$\mathcal{L}[x(t) * h(t)] = \int_0^\infty h(\tau)\left[\int_0^\infty x(t - \tau)e^{-st}dt\right]d\tau$$

$$\tag{7.30}$$

Let $\lambda = t - \lambda$ so that $t = \tau + \lambda$ and $dt = d\lambda$

$$\mathcal{L}[x(t) * h(t)] = \int_0^\infty h(\tau)\left[\int_0^\infty x(\lambda)e^{-s(t+\lambda)}d\lambda\right]d\tau$$

$$= \int_0^\infty h(\tau)e^{-st}d\tau \int_0^\infty x(\lambda)e^{-s\lambda}d\lambda = H(s)X(s)$$

$$\tag{7.31}$$

The Convolution property of a Laplace transform turns into a product of transforms.

**Example**

i)      Consider $g(t) = u(t - 3) - u(t - 4)$.

By applying the Laplace transform on both sides
$$G(s) = \frac{1}{s}(e^{-3s} - e^{-4s})$$
Considering the convolution of $g(t)*g(t)$, which is $G^2(s)$,
$$G^2(s) = \frac{1}{s^2}(e^{-3s} - e^{-4s})^2 = \frac{1}{s^2}(e^{-6s} - 2e^{-7s} + e^{-8s}) \tag{7.32}$$

181

By applying inverse Laplace transform, we get

x(t) = g(t)*g(t) = (t-6)u(t-6) − 2(t-7)u(t-7) + (t-8)u(t-8)

$$= r(t-6) - 2r(t-7) + r(t-8)$$

(7.33)

**7.2.7 Time Integration**

Let X(s) be the Laplace transform of the signal x(t). Then Laplace transform of the integral is given as

$$L\left[\int_0^t x(t)dt\right] = \frac{1}{s}X(s)$$

Below example describes the Time integration property.

From convolution property, we get

$$L\left[\int_0^t x(\lambda)d\lambda\right] = L\left[\int_0^t x(\lambda)u(t-\lambda)d\lambda\right]$$
$$= L[u(t) * x(t)] = \frac{1}{s}X(s)$$

$$L\left[\int_0^t x(t)dt\right] = \frac{1}{s}X(s)$$

(7.34)

**Examples**

i)      Consider $x(t) = u(t)$.
         We know that $X(s) = 1/s$.
         By applying the time integration property,

$$L\left[\int_0^t u(t)dt\right] = L[t] = \frac{1}{s}\frac{1}{s}$$

ii)     Consider $r(t) = t.u(t)$
         We know that $L[t] = \frac{1}{s^2}$
         By applying the time integration property,

182

$$L\left[\int_0^t t\, dt\right] = L\left[\frac{t^2}{2}\right] = \frac{1}{s}\frac{1}{s^2}$$

Similarly, $L[t^2] = \frac{2}{s^3}$

$$L\left[\int_0^t t^2 dt\right] = L[t^2] = \frac{2}{s^3}$$
$$\text{or } L[t^3] = \frac{6}{s^4} = \frac{3!}{s^4}$$

(7.35)

Continuing this the process leads to
$$L[t^n] = \frac{n!}{s^{n+1}}$$

(7.36)

### 7.2.8 Frequency Differentiation

Let X(s) be the Laplace transform of the signal x(t). Then Laplace transform of the Frequency differentiation is given as

$$\mathcal{L}[tx(t)] = -\frac{dX(s)}{ds}$$

Below example describes the Frequency differentiation property, X(s) is given as

$$X(s) = \int_{0^-}^{\infty} x(t)e^{-st}dt$$

Differentiating on both sides, we get

$$\frac{dX(s)}{ds} = \int_{0^-}^{\infty} x(t)(-te^{-st})dt = \int_{0^-}^{\infty} (-tx(t))e^{-st}dt$$
$$= \mathcal{L}[-tx(t)]$$

The Frequency differentiation property is given as

$$\mathcal{L}[tx(t)] = -\frac{dX(s)}{ds}$$

(7.37)

183

Similarly the general form of frequency differentiation property is given as

$$L[t^n x(t)] = (-1)^n \frac{d^n X(s)}{ds^n}$$

(7.38)

**Example**

$$L(t \cos \omega t) = -\frac{d}{ds}\left[\frac{s}{s^2+\omega^2}\right] = \frac{s^2-\omega^2}{(s^2+\omega^2)^2}$$

(7.39)

$$L(t \sin \omega t) = -\frac{d}{ds}\left[\frac{\omega}{s^2+\omega^2}\right] = \frac{2\omega s}{(s^2+\omega^2)^2}$$

(7.40)

### 7.2.9 Time Periodicity

Consider a periodic signal x(t) and it can be expressed as a sum of time-shifted functions as

$$x(t) = x_1(t) + x_2(t) + x_3(t) + L$$

$$= x_1(t) + x_1(t-T)u(t-T) + x_1(t-2T)u(t-2T) + L$$

(7.41)

where 0 < t < T.

By applying the Laplace transform and from the time-shifting property, we get

$$X(s) = X_1(s) + X_1(s)e^{-Ts} + X_1(s)e^{-2Ts} + X_1(s)e^{-3Ts} + L$$

$$= X_1(s)[1 + e^{-Ts} + e^{-2Ts} + e^{-3Ts} + L]$$

(7.42)

But

$$1 + a + a^2 + a^3 + L = \frac{1}{1-a}$$

184

if |a| < 1.

$$X(s) = \frac{X_1(s)}{1 - e^{-Ts}}$$

**Example**

Consider $x_1(t) = \begin{cases} \sin \pi t, & 0 < t < 1 \\ 0, & 1 < t < 2 \end{cases}$

(7.45)

T=2 and $X_1(s) = \frac{\pi}{s^2 + \pi^2}$

(7.46)

The Laplace transform of the periodic signal is

$$X(s) = \frac{X_1(s)}{1 - e^{-Ts}} = \frac{\pi}{(1 - e^{-2s})(s^2 + \pi^2)}$$

(7.47)

### 7.2.10 Modulation

Let X(s) be the Laplace transform of the signal x(t), then for any real number ω,

$$\mathcal{L}[x(t)\cos\omega t] = \frac{1}{2}[X(s + j\omega) + X(s - j\omega)]$$

$$\mathcal{L}[x(t)\sin\omega t] = \frac{j}{2}[X(s + j\omega) - X(s - j\omega)]$$

(7.48)

The Laplace transform of x(t)cosωt and x(t)sinωt are

$$x(t)\cos\omega t = \frac{1}{2}x(t)[e^{-j\omega t} + e^{j\omega t}]$$

185

$$x(t)\sin\omega t = \frac{1}{2j}x(t)[e^{-j\omega t} - e^{j\omega t}]$$

(7.49)

From Equation (7.21),

$$\mathcal{L}[e^{\pm j\omega t}x(t)] = X(s\mp j\omega)$$

(7.50)

Example:

We know that $u(t) \Leftrightarrow \frac{1}{s}$

$$u(t)\cos(\omega_o t) \Leftrightarrow \frac{1}{2}\left[\frac{1}{s+j\omega_o} + \frac{1}{s-j\omega_o}\right] =$$

$$\frac{s}{s^2+\omega_o^2}u(t)\sin(\omega_o t) \Leftrightarrow \frac{1}{2j}\left[\frac{1}{s+j\omega_o} - \frac{1}{s-j\omega_o}\right] = \frac{\omega_o}{s^2+\omega_o^2}$$

(7.51)

### 7.2.11 Initial and Final Values

The initial value x(0) and final value x(∞) are used to relate frequency domain expressions to the time domain as time approaches zero and infinity respectively. These properties can be derived by using the differentiation property as

$$sX(s) - x(0^-) = \mathcal{L}\left[\frac{dx}{dt}\right] = \int_{0^-}^{\infty}\frac{dx}{dt}e^{-st}dt$$

(7.52)

By taking limits, we get

$$\lim_{s\to\infty}[sX(s) - x(0)] = 0$$

Since x(0) is not dependent on s, and it is given as

$$x(0) = \lim_{s\to\infty}sX(s)$$

(7.53)

This is called the initial-value theorem. x(0) can be found using X(s), without using the inverse transform of x(t).

186

If we consider $s \to 0$, then

$$sX(s) - x(0^-) = \int_{0^-}^{\infty} \frac{dx}{dt} e^{-0t} \, dt = \int_{0^-}^{\infty} dx$$

$$= x(\infty) - x(0^-)$$

$$x(\infty) = \lim_{s \to 0} sX(s) \tag{7.54}$$

This is known as the final-value theorem. In the final-value theorem, all poles of X(s) must be located in the left half of the s-plane.

**Example**

$$x(t) = 4 + e^{-2t} \cos 3t \, u(t) \quad \Leftrightarrow \quad X(s) = \frac{4}{s} + \frac{s+2}{(s+2)^2 + 3^2} \tag{7.55}$$

By applying the initial-value theorem,

$$x(0) = \lim_{s \to \infty} sX(s) = \lim_{s \to \infty} 4 + \frac{s^2 + 2s}{s^2 + 4s + 13}$$

$$= \lim_{s \to \infty} 4 + \frac{1 + 2/s}{1 + 4/s + 13/s^2} = 4 + \frac{1+0}{1+0+0} = 5$$

$$x(0) = 5$$

**Example**

i) $\quad x(t) = (5 + e^{-2t}) u(t) \quad \Leftrightarrow \quad X(s) = \frac{5}{s} + \frac{1}{s+2}$ $\tag{7.56}$

Using the final-value theorem,

$$x(\infty) = \lim_{s \to 0} sX(s) = \lim_{s \to 0} \left(5 + \frac{s}{s+2}\right) = 5 + 0 = 5 \tag{7.57}$$

ii) $\quad x(t) = \cos 2t \, u(t) \quad \Leftrightarrow \quad X(s) = \frac{s}{s^2+4}$ $\tag{7.58}$

187

$$x(\infty) = \lim_{s \to 0} sX(s) = \lim_{s \to 0} \frac{s^2}{s^2+2} = 0$$

$$(7.59)$$

The final-value theorem cannot be used to find the value of x(t) = cos2 t, because X(s) has poles at $s = \pm j\sqrt{2}$, which are not in the left half of the s plane.


## 7.3 The Inverse Laplace Transform

It the previous section we have seen the method for finding the Laplace transform. In this section we discuss the method for reversing the process of the previous section and more precisely we reconstruct a function f(t) whose Laplace transform F(s) is given.

**Definition:** Let f(t) be a function such that L (f(t)) = F(s), then f(t) is called the inverse Laplace transform of F(s). The inverse Laplace transform is designated L⁻¹ and we write

$$f(t) = L^{-1}\{F(s)\}.$$

In order to find an inverse transform we must be familiar with the formulas for finding the Laplace transform, see Table 7.1. One should learn to use this table in reverse. However in general the given Laplace transform will not be in the form the allows direct use of the table, so the given F(s) have to be algebraically manipulated in a form that can be found in the table. the most relevant result for this purpose is the linearity property of the inverse Laplace transform which states that

$$L^{-1}\{c_1 F_1(s) + c_2 F_2(s)\} = c_1 L^{-1}\{F_1(s)\} + c_2 L^{-1}\{F_2(s)\}$$

where $c_1$ and $c_2$ are constants.

The proof of this result follows from the definition of the inverse Laplace transform and the corresponding linearity of the Laplace transform.

**Example 7.4** Find

(i) $\quad L^{-1}\left\{\frac{1}{s+2}\right\}$

(ii) $\quad L^{-1}\left\{\frac{1}{s^2+2}\right\}$

188

(iii)    $L^{-1}\left\{\dfrac{-2s+6}{(s^2+4)}\right\}$

(iv)    $L^{-1}\left\{\dfrac{s+5}{(s^2-2s-3)}\right\}$

(v)    $L^{-1}\left\{\dfrac{s+1}{(s^2-4s)}\right\}$

**Solution** (i)    From Table 7.1 $L(e^{at})=\dfrac{1}{s-a}$ Choosing a = -2 we get L

$(e^{-2t})=\dfrac{1}{s+2}$ and consequently by the definition of the inverse Laplace transform

$$L^{-1}\left\{\dfrac{1}{s+2}\right\}=e^{-2t}$$

(ii)    By Table 7.1 for $k=\sqrt{2}$ and the linearity of the inverse Laplace transform we get

$$L^{-1}\left\{\dfrac{1}{s^2+2}\right\}=\dfrac{1}{\sqrt{2}}L^{-1}\left\{\dfrac{\sqrt{2}}{s^2+2}\right\}$$
$$=\dfrac{1}{\sqrt{2}}\sin\sqrt{2}\,t$$

(i)    Solution $L^{-1}\left\{\dfrac{-2s+6}{s^2+4}\right\}=L^{-1}\left\{\dfrac{-2s}{s^2+4}+\dfrac{6}{s^2+4}\right\}$

$$=-2\,L^{-1}\left\{\dfrac{s}{s^2+4}\right\}+6\,L^{-1}\left\{\dfrac{1}{s^2+4}\right\}$$

By the Linearity of the inverse transform,

$$=-2\cos 2t+\dfrac{6}{2}\,L^{-1}\left\{\dfrac{2}{s^2+4}\right\}$$
$$=-2\cos 2t+3\sin 2t \qquad\text{by Table 7.1 (6 and 7)}$$

(ii)    Since $s^2-2s-3=(s-3)(s+1)$ we get

$$\dfrac{s+5}{s^2-2s-3}=\dfrac{s+5}{(s-3)(s+1)}=\dfrac{A}{s-3}+\dfrac{B}{s+1}$$

where A and B are constants to be determined.

$$\dfrac{A}{s-3}+\dfrac{B}{s+1}=\dfrac{A(s+1)+(s-3)B}{(s-3)(s+1)}$$
$$=\dfrac{(A+B)s+(A-3B)}{(s-3)(s+1)}$$

We can write

$$\dfrac{s+5}{(s-3)(s+1)}=\dfrac{(A+B)s+(A-3B)}{(s-3)(s+1)}$$

This implies that

$$s+5=(A+B)s+(A-3B)$$

This gives $A+B=l$ and $A-3B=5$

Subtracting second from first we get B= -1 . Putting this value in the first equation we get A = 2. Therefore, we have

$$L^{-1}\left\{\dfrac{s+5}{s^2-2s-3}\right\}=L^{-1}\left\{\dfrac{2}{s-3}+\dfrac{-1}{s+1}\right\}$$

189

$$= 2\,L^{-1}\left\{\frac{1}{s-3}\right\} - L^{-1}\left\{\frac{1}{s+1}\right\} \quad \text{using linearity of } L^{-1}$$

By Table 7.1 (series no.10, for a = 3 and a = -1) we get

$$L^{-1}\left\{\frac{1}{s-3}\right\} = e^{3t} \text{ and } L^{-1}\left\{\frac{1}{s+1}\right\} = e^{-t}$$

Hence

$$L^{-1}\left\{\frac{s+5}{s^2-2s-3}\right\} = 2\,e^{3t} - e^{-t}$$

(v) $\quad L^{-1}\left\{\frac{s+1}{s^2-4s}\right\} = L^{-1}\left\{-\frac{1}{4}\frac{1}{s} + \frac{5}{4}\frac{1}{s-4}\right\}$

$$= -\frac{1}{4}L^{-1}\left\{\frac{1}{s}\right\} + \frac{5}{4}L^{-1}\left\{\frac{1}{s-4}\right\}$$

$$= -\frac{1}{4}.1 + \frac{5}{4}e^{4t}$$

## 7.4 Shifting Theorems and Derivative of the Laplace Transform

The following theorems are called the shifting theorems.

**Theorem :** (The First Shifting Theorem): Let L (f(t)) = F(s).

Then L $\{e^{at}f(t)\} = F\ (s\text{-}a)$

**Proof :** By definition of L $\{e^{at}f(t)\}$, we write

$$L\ \{e^{at}f(t)\} = \int_0^\infty e^{-st}[e^{at}f(t)]\,dt$$
$$= \int_0^\infty e^{-(s-a)t}f(t)\,dt = F(s-a).$$

**Theorem :** (The second shifting theorem). Let $e^{at}f(t)$ (f(t)) = F(s)

Then L $\{H(t-a)f(t-a)\} = e^{\text{-as}}\ F(s)$

where H is the Heaviside function defined as

$$H(t) = \begin{cases} 0\,if\,t < 0 \\ 1\,if\,t \geq 0 \end{cases}$$

**Proof** L $\{H(t-a)f(t-a)\} = \int_0^\infty e^{-st}H(t-a)f(t-a)\,dt$

$$= \int_a^\infty e^{-st}f(t-a)\,dt$$

because H(t-a) = 0 for t < a and H(t-a) = 1 for t ≥ a. Now let u = t-a in the last integral. We get

$$L\ \{H(t-a)f(t-a)\} = \int_0^\infty e^{-s(a+u)}f(u)\,du$$
$$= e^{-as}\int_0^\infty e^{-st}f(u)\,du$$

190

$$= e^{-as} L (f(u)) = e^{-as} F(s).$$

**Example 7.5** Apply the first shifting theorem to find

(a) $L \{e^{3t} \sin t\}$

(b) $L \{e^{-t} g(t)\}$, where

$$g(t) = \begin{cases} 2t, 0 \le t < 3 \\ -1, t \ge 3 \end{cases}$$

(c) $L^{-1}\left\{\dfrac{4}{s^2-4s+20}\right\}$

**Solution (a)** Since $L \{\sin t\} = \dfrac{1}{s^2+1}$, it follows that

by $L\{e^{3t} \sin \quad t\} = \dfrac{1}{(s-3)^2+1}$

(b) $L\{e^{-t} g(t)\} = F (s-a)$.

where $L (g(t)) = F(s)$.

$F(s) = \int_0^\infty e^{-st} g(t)dt = \int_0^3 e^{-st} g(t)dt + \int_3^\infty e^{-st} g (t)dt$

$= \int_0^3 e^{-st} 2t dt - \int_3^\infty e^{-st} dt$

$= 2\left[e^{-st}(-\frac{1}{s})t\right]_0^3 - 2\int_0^3 e^{-st}(-\frac{1}{s})dt - \left[-\frac{1}{s}e^{-st}\right]_3^\infty$

$= \dfrac{2}{s^2} - \dfrac{2e^{-3s}}{s^2} - \dfrac{7e^{-3s}}{s}$

(c) We have $\dfrac{4}{s^2+4s+20} = \dfrac{4}{(s+2)^2+6}$

$F(s+2) = \dfrac{4}{(s+2)^2+16}$

This means we should choose

$F(s) = \dfrac{4}{s^2+16}$

By the first shifting theorem

$L \{e^{-2t} \sin 4t\} = F(s-(2)) = F(s+2)$

$= \dfrac{4}{(s+2)^2+16}$

and therefore

$L^{-1}\left\{\dfrac{4}{(s+2)^2+16}\right\} = e^{-2t} \sin (4t)$.

**Example 7.6** Compute $L^{-1}\left\{\dfrac{se^{-3s}}{s^2+4}\right\}$.

**Solution:**

$L \{H(t-a)f(t-a)\} = e^{-as}F(s)$

or $H(t-a)f(t-a) = L^{-1}\{e^{-as}F(s)\}$

$F(s) = \dfrac{s}{s^2+4}$

$L^{-1}(F(s)) = L^{-1}\left(\dfrac{s}{s^2+4}\right)$ implies that $f(t) = \cos (2t)$.

Therefore,

$L^{-1}\left\{\dfrac{se^{-3s}}{s^2+4}\right\} = H(t-3)\cos (2(t-3))$.

## 7.5 Derivative of the Laplace Transform

**Theorem :** Let f(t) be piecewise continuous and of exponential order over each finite interval, and let
$$L\,(f(t)) = F(s).$$
Then $F(s)$ is differentiable and
$$F'(s) \;=\; L\,\{-tf(t)\}.$$
**Proof:** Suppose that $|f(t)| \leq Me^{at}$, t>0 and take any $s_0$ >a. Then consider
$$\frac{\partial}{\partial s}\,[e^{-st}f(t)] = \text{-t f(t)}e^{\text{-st}}.$$
Choose $\varepsilon > 0$ such that $s_0 > a + \varepsilon$. Then we have $|t| < e^{\varepsilon t}$ for all t large enough since in fact
$$\lim_{t \to \infty}\left(\frac{s}{e^{et}}\right) = 0.$$
Thus $|\,tf(t)\,| \leq M\,e^{(a+\varepsilon)\,t}$
for all large t and we find that t f(t) is also of exponential order and
$$\int_0^\infty tf(t)e^{-st}\,dt$$
exists by Theorem 7.1, that is, the integral converges uniformly. Hence F(s) is differentiable at $s_0$ and that
$$F'\,(s_0) = \int_0^\infty \frac{\partial}{\partial s}\,[e^{-st}f(t)]dt \text{ at } s{=}s_0.$$
Therefore
$$F'(s) = \text{-}\int_0^\infty tf(t)e^{-st}\,dt$$
$$= \text{L(-t f(t))} \qquad\qquad \text{for all s > a}$$

## 7.6 Transforms of Derivatives, Integrals and Convolution Theorem

### 7.6.1 Transforms of Derivatives and Integrals

The Laplace transform of the derivatives of a differentiable function exist under appropriate conditions. In this section we discuss results that are quite useful in solving differential equations. For solving 2nd order differential equations we need to evaluate the Laplace transforms of $\frac{dy}{dt}$ and $\frac{d^2y}{dt^2}$.

Let f(t) be differentiable for $t \geq 0$ and let its derivative f'(t) be continuous, then by applying the formula for integration by parts we find that

$$L\{f'(t)\} = sF(s) - f(0)$$
(7.60)

$$L\{f'(t) = \int_0^\infty e^{-st} f'(t)dt$$
$$= [e^{-st}f'(t)dt]_0^\infty + s\int_0^\infty e^{-st} f(t)dt$$
$$= -f(0) + sL(f(t))$$
$$= sF(s) - f(0)$$

Here we have used the fact that

$$\lim_{t \to \infty} e^{-st} f(t) = 0$$

Similarly for a twice differentiable function f(t) such that f"(t) is continuous we can prove that

$$L\{f"(t)\} = s2 F(s) - s f(0) - f'(0) \qquad (7.61)$$

In fact we can prove the following theorem, repeated by applying integration by parts.

**Theorem 7.5** Let f,f', -- - - $f^{(n-1)}$ be continuous on $[0,\infty)$ and of exponential order and if $f^{(n)}$ (t) be piecewise continuous on $[0,\infty)$, then

$$L\{f^{(n-1)}(t)\} = s^n F(s) - s^{n-1}f(0) - s^{n-2}(0)\ldots\ldots -f^{(n-1)}(0)$$
(7.62)

where $F(s) = L\{f(t)\}$.

**Theorem:** Let f be piecewise continuous and of exponential order for $t \geq 0$, then

$$L\left\{\int_0^t f(u)du\right\} = \frac{1}{s}L\{f(t)\} = \frac{1}{s}F(s).$$

**Proof:** Let g(t)= $\{\int_0^t$ f(u)du}. Then g'(t) = f(t) and g(0)=0.

Furthermore, g(t) is of exponential order. By Theorem 7.5 L {g'(t)} = s L {g(t)} -g(0)

  or  L {f(t)} = sL $\{\int_0^t f$ (u) du}

  or L $\left\{\int_0^t f(u)du\right\} = \frac{1}{s}L\{f(t)\}$

**Example 7.7** (a) Using the Laplace transform of f" find L {sin k t }.

 (b) Show that L $^{-1}\left\{\frac{1}{s}F(s)\right\} = \int_0^t f(u)du.$

**Solution** (a) Let $f(t) = \sin k\, t$, then $f'(t) = k \cos k\, t$, $f''(t) = -k^2 \sin k\, t$, $f(0)=0$ and $f'(0)=k$.

Therefore  $L\{f''(t)\} = s^2\, F(s) - sf(0) - f'(0)$

or $L\{f''(t)\} = s^2\, F(s) - k$

or $L\{-k^2 \sin \quad kt\} = s^2\, F(s) - k$

where $F(s) = L\{f(t)\} = L\{\sin k\, t\}$

$L\{-k^2 \sin k\, t\} = s^2\, F(s) - k = s^2\, L\{\sin k\, t\} - k$. Solving for $L\{\sin k\, t\}$ we get

$L\{\sin k\, t\} = \dfrac{k}{s^2+k^2}$

$$\left[\begin{array}{l} -k^2\, L\{\sin kt\} = s^2\, L\{\sin kt\} - k \text{ or} L\{\sin kt\}\{s^2 + k^2\} = k \\ \text{or } L\{\sin kt\} = \dfrac{k}{s^2 + k^2} \end{array}\right]$$

(b)    By Theorem 7.6

$$L\left\{\int_0^t f(u)ds\right\} = \frac{1}{s}F(s)$$

This implies that $\int_0^t f(u)du = L^{-1}\left\{\frac{1}{s}F(s)\right\}$

## 7.6.2 Convolution

**Definition :**(Convolution). Let f and g be piecewise continuous functions for $t \geq 0$. Then the **convolution** of f and g denoted by f*g, is defined by the integral

$(f*g)\,(t) = \int_0^t f(u)g(t-u)du$

$= \int_0^t g(u)f(t-u)du$

$= (g*f)(t).$

**Theorem :**(Convolution theorem). Let f and g be piecewise continuous and of exponential order for $t \geq 0$, then the Laplace transform of f * g is given by the product of the Laplace transform of f and the Laplace transform of g. That is

$L\{f * g\} = F\,(s)\,G(s).$

**Proof :**   Let $F = L\,(f)$ and $G = L\,(g)$. Then

$F(s)G(s) = F(s) \int_0^\infty e^{-st}\, g(t)dt$

$= \int_0^\infty F(s)e^{-su}\, g(u)du$

in which we changed variable of integration from t to u and brought $F(s)$ inside the integral

Let us recall that $e^{-su}\, F(s) = L\,\{H\,(t-u)\, f(t-u)\}$

where F(s) = L {f(t)} and H (.) is the Heaviside function, see Theorem 7.3.

Substitute this into the integral for F(s)G(s) to get

$$F(s)G(s) = \int_0^\infty L\{H(t-u)f(t-u)\}g(u)du$$
(7.63)

But, from the definition of the Laplace transform,

$$L\{H(t-u)f(t-u)\} = \int_0^\infty e^{-st}\{H(t-u)f(t-u)\}dt$$

Substituting this into (7.5) we get

$$F(s)G(s) = \int_0^\infty \left[\int_0^\infty e^{-st}H(t-u)f(t-u)dt\right]g(u)du$$

$$= \int_0^\infty \int_0^\infty e^{-st}g(u)H(t-u)f(t-u)dtdu$$

Let us recall that $H(t-u) = 0$ if $0 \le t < u$
$$= 1 \text{ if } t \ge u$$

Therefore,

$$F(s)\,G(s) = \int_0^\infty \int_u^\infty e^{-st}g(u)f(t-u)dtdu.$$

The Laplace integral is over shaded region, consisting of points satisfying $0 \le u \le t < \infty$. Reversing the order of integration gives us

$$F(s)G(s) = \int_0^\infty \int_0^t e^{-st}g(u)f(t-u)dudt.$$

$$= \int_0^\infty e^{-st}\left[\int_0^t g(u)f(t-u)du\right]dt$$

$$= \int_0^\infty e^{-st}(f*g)(t)dt$$

$$= L\{f*g\}.$$

It follows immediately from Theorem 7.7 that

**Theorem:** Let $L^{-1}(F)=f$, $L^{-1}(G)=g$. Then

$$L^{-1}\{F\,G\} = f*g$$

**Example 7.8.** Evaluate $L^{-1}\left\{\dfrac{1}{(s^2+k^2)^2}\right\}$

**Solution:** Let $F(s) = G(s) = \dfrac{1}{s^2+k^2}$

so that $f(t) = g(t) = \dfrac{1}{k}L^{-1}\left\{\dfrac{1}{s^2+k^2}\right\} = \dfrac{1}{k}\sin kt$

By Convolution Theorem

$$L^{-1}\left\{\dfrac{1}{(s^2+k^2)^2}\right\} = \dfrac{1}{k^2}\int_0^t \sin k\,u\,\sin k\,(t-u)du$$

$$= \dfrac{1}{2k^2}\int_0^t[\cos k(2u-t) - \cos kt]du$$

$$= \dfrac{1}{2k^2}\left[\dfrac{1}{2k}\sin k\,(2u-t) - u\cos kt\right]_0^t$$

$$= \dfrac{\sin kt - kt\cos kt}{2k^3}$$

195

### 7.6.3 Unit Impulse and the Dirac Delta Function

Very often one encounters the concept of an impulse, which may be thought as a force of large magnitude applied over an instant of time. Impulse can be defined as follows.

For any positive number $\varepsilon$, the pulse $\delta_\varepsilon$ is defined by
$$\delta\varepsilon\,(t) \ =\frac{1}{\varepsilon}\,[\,H(t) - H\,(t - \varepsilon)\,]$$
This is a pulse of magnitude $\frac{1}{\varepsilon}$ and duration $\varepsilon$.

By allowing $\varepsilon$ to approach zero, we obtain pulses of increasing magnitude over shorter time intervals.

Dirac's delta function is understood as a pulse of infinite magnitude over an infinitely short duration and is defined to be
$$\delta(t) = \lim_{\varepsilon \to 0+} \delta_\varepsilon(t).$$
It may be observed that it is not a function in the conventional sense but it is a more general object called **distribution.** Nevertheless, for historic reason it continues to be called the delta function. It is named for the Nobel laureate physicist P.A.M. Dirac who was also the guide and mentor of another Nobel laureate physicist Abdul Salam–founder Director of the International Centre of Theoretical physics, Trieste, Italy. The shifted delta function $\delta(t-a)$ is zero except for t=a, where it has its infinite spike. It is interesting to note that the Laplace transform of the Dirac delta function $\delta(t)$; that is, L $\{\delta(t)\} = 1$.

**Verification:**
$$L\{\,\delta_\varepsilon(t\text{-}a)\,\} =\frac{1}{\varepsilon}\left[\int_0^\infty e^{-st}(H(t - a) - H(t - a - \varepsilon))dt\right]$$
$$=\frac{1}{\varepsilon}\left[\frac{1}{s}e^{-as} -\frac{1}{s}e^{-(a+\varepsilon)s}\right]$$
$$=\frac{e^{-as}(1-e^{-\varepsilon s})}{\varepsilon s}$$

This suggests that we define
$$L\,(\{\delta(t\text{-}a)\} = \lim_{\varepsilon \to 0+} \frac{e^{-as}(1-e^{-\varepsilon s})}{\varepsilon s}$$
$$=e^{\text{-}as}$$

In particular choose $a = 0$ we get
$$L\,(\delta\,(t))= 1$$
The following result is known as the filtering property.

**Theorem 7.9 (Filtering Property) Let** a>0 and let f be integrable on $[0,\infty)$ and continuous at a. Then

$$\int_0^\infty f(t)\delta(t-a)dt = f(a).$$

**Proof:** is straight forward and is obtained by putting values of $\delta_\varepsilon$ and taking limit as $\varepsilon \to 0+$.

If we apply the filtering property to f(t)= e$^{-st}$, we get

$$\int_0^\infty e^{-st}\delta(t-a)dt = e^{-as}.$$

Furthermore, if we change notation in the filtering property and write it as

$$\int_0^\infty f(u)\delta(u-t)du = f(t),$$

then we can recognize the convolution of $f$ with $\delta$ and read the last equation as

$$f*\delta = f$$

**The delta function therefore acts** as an identity for the product defined by the convolution of two functions. (The convolution defined earlier is treated as a special type of product. The Dirac delta function is its identity).

**Theorem :** Transform of a periodic function

If f(t) is piecewise continuous on $[0,\infty)$, of exponential order and periodic with period T, $(f(t+T) = f(t))$ then $L\{f(t)\} = \frac{1}{1-e^{-sT}}\int_0^T e^{-st}f(t)dt$

**Proof:** Writing the Laplace transform of f as:

$$L\{f(t)\} = \int_0^T e^{-st}f(t)\,dt + \int_T^\infty e^{-st}f(t)\,dt$$

Letting t= u+T in the last integral

$$\int_T^\infty e^{-st}f(t)\,dt = \int_0^\infty e^{-(u+T)s}f(u+T)\,dt$$

$$= e^{-sT}\int_0^\infty e^{-su}f(u)du = e^{-sT}L\{f(t)\}$$

Therefore L $\{f(t)\} = \int_0^T e^{-st}f(t)dt + e^{-sT}L\{f(t)\}$

Thus L $\{f(t)\} = \frac{1}{1-e^{-sT}}\int_0^T e^{-st}f(t)dt$

**Example 7.9:** Find the Laplace Transform of the square wave function E(t) of period T=2 defined as E(t) = $\begin{cases} 1,0 \le t < 1 \\ 0,1 \le t < 2 \end{cases}$

**Solution:** L $\{E(t)\} = \frac{1}{1-e^{-2s}}\int_0^2 e^{-st}E(t)dt$

$$= \frac{1}{1-e^{-2s}}\left[\int_0^1 e^{-st}1dt + \int_1^2 e^{-st}.0dt\right]$$

$$= \frac{1}{1-e^{-2s}}\frac{1-e^{-s}}{s}$$

$$= \frac{1}{s(1+e^{-s})} \qquad \text{(using 1-e}^{-2s} = (1+e^{-s})(1-e^{-s}).$$

## 7.7 Applications to Differential and Integral Equations

In this section we discuss applications of the Laplace transform and related methods in finding solutions of differential equations with initial conditions and integral equations.

Essentially Laplace transform converts initial value problem to an algebraic problem, incorporating initial conditions into the algebraic manipulations. There are three basic steps:

(i)    Take the Laplace transform of both sides of the given differential equation, making use of the linearity property of the transform.

(ii)   Solve the transformed equation for the Laplace transform of the solution function.

(iii)  Find the inverse transform of the expression F(s) found in step (ii).

**Example 7.10** Apply the Laplace transform to solve the initial-value problem

$$y'+2y = 0, y(0) = 1.$$

**Solution:** Given equation is y'+2y=0. Taking the Laplace transform of both sides of this equation yields

$$L\{y'+2y\} = L\{0\}$$
$$\text{or } L\{y'\}+2 L(y) = L\{0\} \text{ by the linearity of the Laplace}$$
transform.

Let L {y(t)} = Y(s) and applying equation (7.2), the previous equation takes the form

$$s Y(s)-1+2 Y(s)=0$$

Solving for Y(s) we have

$$Y(s) = \frac{1}{s+2}$$

The function y(t) is then found by taking the inverse transform of this equation. Thus

$$y(t)= L^{-1}\left(\frac{1}{s+2}\right)= e^{-2t}, \text{ see Table 7.1}$$

y(t) = e$^{-2t}$ is the solution of the given initial value problem.

**Example 7.11** Apply the Laplace transform to solve the initial-value problem

$$y'-4y=1, y(0)=1.$$

**Solution:** Let L $\{y(t)\}=Y(s)$

Taking the Laplace transform of the differential equation, using the linearity of L and equation (7.2) we get

$$L\{y' - 4y\} = L\{1\}$$

or $\quad L\{y'\} - 4L\{y\} = {}^1/_s$

or $\quad sY(s) - y(0) - 4Y(s) = {}^1/_s$

or $\quad sY(s) - 1 - 4Y(s) = {}^1/_s$

or $\quad Y(s)(s - 4) = 1 + \dfrac{1}{s}$

or $\quad Y(s) = \dfrac{1}{(s-4)} + \dfrac{1}{s(s-4)}$

Taking the inverse Laplace transform of this equation we have

$$L^{-1}\{Y(s)\} = L^{-1}\left\{\frac{1}{s-4} + \frac{1}{s(s-4)}\right\}$$

By the linearity of L$^{-1}$ we get

$$L^{-1}\{Y(s)\} = L^{-1}\left\{\frac{1}{s-4}\right\} + L^{-1}\left\{\frac{1}{s(s-4)}\right\}$$

$$L^{-1}\left\{\frac{1}{s-4}\right\} = e^{4t}, L^{-1}\left\{\frac{1}{s(s-4)}\right\} = \frac{1}{4}L^{-1}\left\{\frac{1}{s-4} - \frac{1}{s}\right\}$$

$$= \frac{1}{4}L^{-1}\left\{\frac{1}{s-4}\right\} - \frac{1}{4}L^{-1}\left\{\frac{1}{s}\right\}$$

$$= \frac{1}{4}e^{4t} - \frac{1}{4}$$

Thus

$$y(t) = e^{4t} + \frac{1}{4}e^{4t} - \frac{1}{4}$$

$$= \frac{5}{4}e^{4t} - \frac{1}{4}$$

is the solution of the given initial value problem.

**Example 7.12**      Solve $y'' + 4y = e^{-t}, y(0) = 2, y'(0) = 1$

        **Solution:** Let $Y(s) = L\,y((t))$. Taking the Laplace transform of both sides of the given differential equation we get

$$L\{y''\} + 4L\{y\} = L(e^{-t})$$

By equation (7.3) $L\{y''\}=s^2Y(s)-2s-1$ keeping in mind the given initial conditions      and         so             $s^2$

$$Y(s) - 2s - 1 + 4Y(s) = \frac{1}{s+1}$$

Solving for Y(s) we get

$$Y(s) = \frac{1}{s^2 + 4}\left[1 + \frac{1}{s+1} + 2s\right]$$

$$= \frac{2s^2 + 3s + 2}{(s+1)(s^2 + 4)}$$

The partial fractions for this are

199

$$\frac{2s^2 + 3s + 2}{(s + 1)(s^2 + 4)} = \frac{A}{s + 1} + \frac{Bs + C}{s^2 + 4}$$

for which $A = \frac{1}{5},B=9/5, and\ C = \frac{6}{5}$. Thus

$$Y(s) = \frac{1}{5(s + 1)} + \frac{9s + 6}{5(s^2 + 4)}$$

$$= \frac{1}{5(s + 1)} + \frac{9s}{5(s^2 + 4)} + \frac{6}{5(s^2 + 4)}$$

Taking the inverse transform yields

$$y(t) = \frac{1}{5}e^{-t} + \frac{9}{5}\cos 2t + \frac{3}{5}\sin 2t$$

**Example 7.13** Solve

$$y'' + 4y' + 3y = e^t$$
$$y(0) = 0, y'(0) = 2$$

**Solution:** Take the Laplace transform of the given differential equation to get

$$L\{y''\} + L\{4y'\} + L\{3y\} = L\{et\}$$

By equations (7.2), (7.3) and applying initial conditions we get
$L\{y''\} = s2\,Y(s) - sy(0) - y'(0) = s2\,Y(s) - 2$ and
$$L\{y'\} = sY\,(s) - y(0) = sY(s)$$

Therefore,

$$s2\,Y(s) - 2 + 4sY(s) + 3Y(s) = \frac{1}{s - 1}$$

Solving this for Y(s) we get

$$Y(s) = \frac{2s - 1}{(s - 1)(s^2 + 4s + 3)}.$$

Let $\frac{2s-1}{(s-1)(s^2+4s+3)} = \frac{A}{s-1} + \frac{B}{s+1} + \frac{C}{s+3}$

This equation can hold only if, for all s,

$A\,(s + 1)\,(s + 3) + B(s - 1)\,(s + 3) + C(s - 1)\,(s + 1)$
$= 2s - 1.$

Now choose values of s to simplify the task of determining A,B, and C. Let s=1 to get 8A =1, so $A=\frac{1}{8}$.Let s = -1 to get -4B= - 3, so B= $^3/_4$.

Choose s= - 3 to get 8C = - 7,so C= $-\frac{7}{8}$.

Then

$$Y(s) = 1/8\ 1/(s - 1) + (3/4)/(s + 1) - (7/8)/(s + 3).$$

By Table 7.1 we find that

$$L-1\{Y(s)\} = \frac{1}{8}L-1\left(\frac{1}{s-1}\right)+\frac{3}{4}L-1\left(\frac{1}{s+1}\right)-\frac{7}{8}L$$
$$-1\left\{\frac{1}{s+3}\right\}$$
$$y(t) = \frac{1}{8}e^t + \frac{3}{4}e^{-t} - \frac{7}{8}e^{-3t}$$

This is the solution of the given initial value problem.

**Example 7.14** Find $L\{f(t) * g(t)\}$, where $f(t) = e^{-t}$ and $g(t) =$ sin $2t$.

**Solution.**    By Theorem 7.7 we have
$$L\{f(t) * g(t)\} = F(s)G(s)$$
where
$$F(s) = \int_0^\infty e^{-st}e^{-t}dt = \frac{1}{s+1}$$

$$G(s) = \left(\frac{2}{s^2+4}\right)$$
Thus $L\{f(t) * g(t)\} = \left(\frac{1}{s+1}\right)\left(\frac{2}{s^2+4}\right)$
$$= \frac{2}{(s+1)(s^2+4)}$$

**Example 7.15** Evaluate $L\left\{\int_0^t e^u \sin \quad (t-u)du\right\}$

**Solution:**    $\int_0^t e^u \sin \quad (t-u)du = f(t) * g(t)$ where $f(t) =$
$e^t$                              and                              $g(t) = \sin t$
by definition of the convolution.
By theorem 7.7 we get
$$L\{f(t) * g(t)\} = L\{f(t)\}L\{g(t)\}$$

$$= \frac{1}{s+1}\frac{1}{s^2+1}$$

$$= \frac{1}{(s-1)(s^2+1)}$$

An equation involving an unknown function $f(t)$, known functions g(t) and h(t) and integral of f and g is called a **Volterra integral equation** for $f(t)$ :

$f(t) = g(t) + \int_0^t f(u)h(t-u)du.$

**Example 7.16** Solve the following Volterra integral equation for $f(t)$:

$$f(t) = 3t^2 - e^{-t} - \int_0^t f(u)e^{t-u}du$$

**Solution:** We identify $h(t - u) = e^{t-u}$ so that $h(t) = e^t$
Take the Laplace transform of both sides, we have

$$L\{f(t)\} = L\{3t^2\} - L\{e^{-t}\} - L\left\{\int_0^t f(u)e^{t-u}du\right\}$$

$$L\{f(t)\} = F(s), L\{3t^2\} = 3L\{t^2\} = 3\frac{2}{s^3}$$

$$L\{e^{-t}\} = \frac{1}{s+1}, L\left\{\int_0^t f(u)e^{t-u}du\right\} = L\{f(t) * h(t)\} =$$
$$L\{f(t)\}L\{h(t)\}$$

$$\left\{\int_0^t f(u)e^{t-u}du\right\} = L\{f(t)\}L\{e^t\}$$

$$= F(s).\frac{1}{s-1}$$

Therefore,

$$F(s) = \frac{6}{s^3} - \frac{1}{s+1} - \frac{1}{s-1}F(s)$$

$$(s)\left[1 + \frac{1}{s-1}\right] = \frac{-s^3 + 6s + 6}{s^3(s+1)}$$

$$F(s)\left[\frac{s}{s-1}\right] = \frac{-s^3 + 6s + 6}{s^3(s+1)}$$

$$F(s) = \frac{6}{s^3} - \frac{6}{s^4} + \frac{1}{s} - \frac{2}{s+1}$$

by carrying out the partial fraction decomposition.
The inverse transform then gives

$$f(t) = 3L - 1\left\{\frac{2!}{s^3}\right\} - L - 1\left\{\frac{3!}{s^4}\right\} + L - 1\left\{\frac{1}{s}\right\} - 2L$$

$$- 1\left\{\frac{1}{s+1}\right\} = 3t^2 - t^3 + 1 - 2e^{-t}$$

**Example 7.17** Find f(t) such that
$$f(t) = 2t^2 + \int_0^t f(t-u)e^{-u}du$$

**Solution:** It is clear that

$$f(t) * g(t) = \int_0^t f(t-u)e^{-u}du$$

and by Theorem 7.7.

$$L\{f(t) * g(t)\} = L\{f(t)\}L\{g(t)\}$$

$$= F(s)\frac{1}{s+1}$$

202

By taking the Laplace transform of both sides of the integral equation we get

$$L\{f(t)\} = L\{2t^2\} + F(s)\frac{1}{s+1}$$

or $\quad F(s) = 2.\frac{2}{s^3} + \frac{1}{s+1}F(s)$

or $\quad F(s)\left[\frac{s}{s+1}\right] = \frac{4}{s^3}$

or $\quad F(s) = \frac{4(s+1)}{s^4}$

$$= \frac{4}{s^3} + \frac{4}{s^4}$$

Taking inverse Laplace transform we get

$$f(t) = 2L^{-1}\left\{\frac{2}{s^3}\right\} + \frac{2}{3}L^{-1}\left(\frac{3!}{s^4}\right) = 2t^2 + \frac{2}{3}t^3$$

**Example 7.18** Find the function $f(t)$ if

$$f(t) = t + \int_0^t f(u)\sin(t-u)du$$

**Solution:** We can identify the integral as
$$f(s)*h(t) \text{ where } h(t) = \sin t$$
Taking the Laplace transform of both sides of the integral equation we get

$$L\{f(t)\} = L\{t\} + L\{f(t)*h(t)\}.$$
$$L\{f(t)*h(t)\} = L\{f(t)\}L\{h(t)\}$$
$$= F(s)\frac{1}{s^2+1}$$

Thus

$$F(s) = \frac{1}{s^2} + F(s)\frac{1}{s^2+1}$$

or $\quad F(s)\frac{s^2}{s^2+1} = \frac{1}{s^2}$

or $\quad F(s) = \frac{s^2+1}{s^4} = \frac{1}{s^2} + \frac{1}{s^4}$

Taking the inverse Laplace transform of this equation we get
$$f(t) = t + \frac{1}{6}t^3$$

**Example 7.19** A spring is attached to a 16-lb block resting on a frictionless plane. A horizontal force of 4 lb is applied to the block through the spring for 3 sec. and then released. Describe the resulting motion if the block is initially at rest and the spring constant is equal to 2.

**Solution:** The differential equation of the system is
$$16/32\, y''(t) + 2\,y(t) = f(t), y(0) = 0, y'(0) = 0$$

203

where the applied force f(t) is given by

$$f(t) = \begin{cases} 4, 0 < t < 3 \\ 0, t > 3 \end{cases}$$

The **unit step function** is denoted by $u_a(t)$ and defined by

$$u_a(t) = \begin{cases} 0, t < a \\ 1, t > a \end{cases}$$

$$f(t) = 4 - 4 u_3(t)$$

The differential equation can be written as

$$1/2\, y''(t) + 2y(t) = 4 - 4\, u_3(t)$$

or

$$y''(t) + 4y(t) = 8 - 8\, u_3(t)$$

Let L $\{y(t)\}$ = Y(s) then

$$s2\, Y(s) + 4Y(s) = \frac{8}{s} - \frac{8}{s} e^{-3s}$$

Therefore

$$Y(s) = \frac{8}{s(s^2 + 4)} - \frac{8}{s(s^2 + 4)} e^{-3s}$$

Using partial fractions expansions,

$$Y(s) = \frac{2}{s} - \frac{2s}{s^2 + 4} - \frac{2}{s} e^{-3s} + \frac{2s}{s^2 + 4} e^{-3s}$$

The appropriate inverse Laplace transforms yield

$$y(t) = 2 - 2\cos 2t - 2\, u_3(t) + 2\cos 2(t - 3)\, u_3(t)$$
$$= 2(1 - \cos 2t) - 2\,[1 - \cos 2(t - 3)]\, u_3(t)$$
$$= \begin{cases} 2(1 - \cos 2t), 0 < t < 3 \\ 2[\cos 2(t - 3) - \cos 2t], t > 3 \end{cases}$$

**Example 7.20** Solve the problem

$$y'' - 2y' - 8y = f(t); \quad y0) = 1, y'(0) = 0.$$

**Solution :** Apply the Laplace transform inserting the initial values, to obtain

$$L\,(y'' - 2y' - 8y) = (s2Y(s) - s) - 2(sY(s) - 1) - 8Y(s)$$
$$= F(s).$$

Then

$$(s2 - 2s - 8)Y(s) - s + 2 = F(s).$$

So

$$Y(s) = \frac{1}{s^2 - 2s - 8} F(s) + \frac{s - 2}{s^2 - 2s - 8}$$

Use a partial fractions decomposition to write

$$Y(s) = \frac{1/6}{s - 4} F(s) - \frac{1/6}{s + 2} F(s) + \frac{1/3}{s - 4} + \frac{2/3}{s + 2}$$

Taking inverse Laplace transform, we get

$$y(t) = \frac{1}{6} e^{4t} f(t) - \frac{1}{6} e^{-2t} f(t) + \frac{1}{3} e^{4t} + \frac{2}{3} e^{-2t}$$

**Example 7.21** Consider the system of differential equations and initial conditions for the functions x and y:

$$x'' - 2x' + 3y' + 2y = 4.$$
$$2y' - x' + 3y = 0.$$
$$x(0) = x'(0) = y(0) = 0. \text{ Solve for } x \text{ and } y.$$

By applying the Laplace transform to the differential equations, incorporating the initial conditions, we get

$$s2X\,(s) - 2sX(s) + 3sY(s) + 2Y(s) = \frac{4}{s}$$

$$2sY(s) - X(s) + 3Y(s) = 0.$$

Solve these equations for X(s) and Y(s) to get

$$X(s) = \frac{4s + 6}{s^2(s + 2)(s - 1)}$$

and

$$Y(s) = \frac{2}{s(s + 2)(s - 1)}$$

A partial fractions decomposition yield

$$X(s) = -\frac{7/2}{s} - 3\frac{1}{s^2} + \frac{1/6}{s + 2} + \frac{10/3}{s - 1}$$

and

$$Y(s) = -\frac{1}{s} + \frac{1/3}{s + 2} + \frac{2/3}{s - 1}.$$

Applying the inverse Laplace transform, we obtain the solution

$$x(t) = -\frac{7}{2} - 3t + \frac{1}{6}e^{-2t} + \frac{10}{3}e^t$$

and

$$y(t) = -1 + \frac{1}{3}e^{-2t} + \frac{2}{3}e^t$$

## Problems

7.1 Find the Laplace transform of the following functions and establish the region of convergence (ROC) for each case.
(a) $e^{2t}\,u(t)$,
(b) $2\delta(t)$.

7.2 Find the Laplace transform of $x(t) = \sin 2.5t\,u(t)$.

7.3 By using linearity property, find $X(s)$
$$x(t) = 2e^{-5t} + 3e^{7t}$$

7.4 Apply the Laplace transform to solve the initial-value problem
$$y' + 3y = 1, y(0) = 1.$$

7.5 Apply the Laplace transform to solve the initial-value problem
$$y'' + 2y = 1, y(0) = 0.1.$$

7.6 Apply the Laplace transform to solve the initial-value problem
$$y'' + 10y = 1, y(0) = 0.1.$$

7.7 Apply the Laplace transform to solve the initial-value problem
$$y'' + 2y = 10, y(0) = 1.$$

# Chapter 8

## Z-Transform

The z-transform is useful for the manipulation of discrete data sequence and has acquired a new significance in the formulation and analysis of discrete-time systems. It is used extensively today in the area of applied mathematics digital signal processing, control theory, population science, economics. These discrete models are solved with difference equations in a manner that is analogous to solving continuous models with differential equations. The role played by the z-transform in the solution of difference equations corresponds to that represented by the Laplace transforms in the solution of differential equations. In this chapter, the z-transform representation will present also the region of convergence (ROC), properties of the z-Transform, and the Inverse z-Transform.

### 8.1 Z-Transform representation

In the previous chapters, we saw that the discrete-time Fourier transform (DTFT) of a sequence $x(n)$ is equal to the sum:

$$X(e^{j\omega}) = \sum_{n=-\infty}^{\infty} x(n)\, e^{-jn\omega}$$

$$(8.1)$$

However, for this series to converge, it is necessary that the signal be summable.

Then, the z-transform is a generalization of the DTFT that allows one to deal with such sequences and is defined as follows:

**Definition:** The z-transform of a discrete-time signal x(n) is characterized by:

$$X(z) = \sum_{n=-\infty}^{\infty} x(n)\, z^{-n}$$

$$(8.2)$$

In other words, the relation between $X(e^{j\omega})$ and $X(z)$ is:

$$X(e^{j\omega}) = X(z)|_{z=e^{j\omega}}$$

$$(8.3)$$

where $z = e^{j\omega}$ is a complex variable.

According to the fact that for most situations, the digital signal $x(n)$ is the causal sequence, that is, $x(n) = 0$ for $n < 0$, then $X(z)$ will be:

$$X(z) = \sum_{n=0}^{\infty} x(n) \, z^{-n}$$

$$(8.4)$$

Thus, the definition in the above Equation referred to as a one-sided z-transform or a unilateral transform.

### Example 8.1

Prove that the convolution in time domain implies convolution in the z-domain, i.e.,

$$Z(x_1(n) * x_2(n)) = Z(x_1(n)) \, Z(x_2(n)) = X_1(z)X_2(z)$$

**Solution:**

According to the definition of the convolution, we have

$$x(n) = x_1(n) * x_2(n) = \sum_{k=0}^{\infty} x_1(n-k)x_2(k)$$

Taking z-transform

$$X(n) = Z(x_1(n) * x_2(n)) = Z\left( \sum_{k=0}^{\infty} x_1(n-k)x_2(k) \right)$$

$$= \sum_{n=0}^{\infty} \left( \sum_{k=0}^{\infty} x_1(n-k)x_2(k) \right) z^{-n}$$

It was written as

$$X(n) = \sum_{k=0}^{\infty} x_2(k) \, z^{-k} . \sum_{n=0}^{\infty} x_1(n-k) \, z^{-(n-k)}$$

Let $n - k = m$, and then the second summation can be written as

$$X(n) = \sum_{k=0}^{\infty} x_2(k)\, z^{-k} \cdot \sum_{m=-k}^{\infty} x_1(m)\, z^{-m}$$

Using causality of both sequences, the second summation can be started from $m = 0$ instead of $m = -k$. Therefore, using the definition of z-transform, we have

$$X(n) = X_1(n) . X_2(n)$$

Hence, proved.

## 8.2 Region of convergence (ROC)

In the above Equation, all the values of z that make the summation to exist (the sum converged) form a region of convergence in the z-transform domain, while all other values of z outside the region of convergence will cause the summation to diverge. The region of convergence is defined based on the particular sequence $x(n)$ Being applied. Note that we deal only with the unilateral z-transform, and hence when performing inverse z-transform (which we shall study later), we are restricted to the causal sequence.

### Example 8.2

Given the sequence $x(n) = u(n)$. Find the z-transform of $x(n)$.

### Solution:

The z-transform is given by

$$X(z) = \sum_{n=0}^{\infty} u(n)\, z^{-n} = \sum_{n=0}^{\infty} (z^{-1})^n$$

It is an infinite geometric series that converges to

$$X(z) = \frac{z}{z-1}$$

with a condition $|z^{-1}| < 1$.

The region of convergence for all values of z is given as $|z| > 1$

**Example 8.3**

Considering the exponential sequence $x(n) = a^n \, u(n)$

Find the z-transform of the sequence $x(n)$.

*Solution*:
  From the definition of the z-transform, it follows that

$$X(z) = \sum_{n=0}^{\infty} a^n \, u(n) \, z^{-n} \;\; = \;\; \sum_{n=0}^{\infty} (az^{-1})^n$$

This geometric series which will converge for $|az^{-1}| < 1$

$$X(z) = \frac{z}{z-a} \;\;, \;\;\; for \;\; |z| > |a|$$

The z-transform for common sequences summarized in Table 8.1.

**Table 8.1 Table of z-transform pairs (for causal sequences)**

| No | Signal $x(n)$, $\quad n \geq 0$ | z-Transform $Z(x(n)) = X(z)$ | Region of Convergence |
|----|------|------|------|
| 1 | $x(n)$ | $\displaystyle\sum_{n=0}^{\infty} x(n) \, z^{-n}$ | |
| 2 | $\delta(n)$ | $1$ | Entire z-plane |
| 3 | $a \, u(n)$ | $\dfrac{az}{z-1}$ | $|z| > 1$ |
| 4 | $n \, u(n)$ | $\dfrac{z}{(z-1)^2}$ | $|z| > 1$ |
| 5 | $n^2 \, u(n)$ | $\dfrac{z(z+1)}{(z-1)^3}$ | $|z| > 1$ |
| 6 | $a^n \, u(n)$ | $\dfrac{z}{z-a}$ | $|z| > |a|$ |

210

| 7 | $e^{-na} u(n)$ | $\dfrac{z}{z - e^{-a}}$ | $|z| > e^{-a}$ |
|---|---|---|---|
| 8 | $n\, a^n u(n)$ | $\dfrac{az}{(z - a)^2}$ | $|z| > |a|$ |
| 9 | $\sin(an)\, u(n)$ | $\dfrac{z \sin(a)}{z^2 - 2z \cos(a) + 1}$ | $|z| > |1|$ |
| 10 | $\cos(an)\, u(n)$ | $\dfrac{z\,(z - \cos(a))}{z^2 - 2z \cos(a) + 1}$ | $|z| > |1|$ |
| 11 | $a^n \sin(bn)\, u(n)$ | $\dfrac{[a \sin(b)]\, z}{z^2 - [2a \cos(b)]z + a^2}$ | $|z| > |a|$ |
| 12 | $a^n \cos(bn)\, u(n)$ | $\dfrac{z\,[z - a \cos(b)]}{z^2 - [2a \cos(b)]z + a^2}$ | $|z| > |a|$ |
| 13 | $e^{-an} \sin(bn)\, u(n)$ | $\dfrac{[e^{-a} \sin(b)]\, z}{z^2 - [2e^{-a} \cos(b)]z + e^{-2a}}$ | $|z| > e^{-a}$ |
| 14 | $e^{-an} \cos(bn)\, u(n)$ | $\dfrac{z\,[z - e^{-a} \cos(b)]}{z^2 - [2e^{-a} \cos(b)]z + e^{-2a}}$ | $|z| > e^{-a}$ |
| 15 | $2|A||P|^n \cos(n\theta + \phi)\, u(n)$ where $P$ and $A$ are complex constants defined by $P = |P|\angle\theta$, $A = |A|\angle\phi$ | $\dfrac{Az}{z - P} + \dfrac{A^* z}{z - P^*}$ | |

## Example 8.4

Find the z-transform for each of the following sequences $x(n)$.

a) $x(n) = 2.5\, u(n)$
b) $x(n) = 2\sin(\dfrac{\pi n}{4})$
c) $x(n) = (0.8)^n u(n)$
d) $x(n) = (0.8)^n \sin(\dfrac{\pi n}{4})\, u(n)$
e) $x(n) = e^{-0.5n} \cos\left(\dfrac{\pi n}{4}\right) u(n)$

Solution

a)      $x(n) = 2.5\, u(n)$

*From table*

$$X(z) = Z[x(n)] = Z[2.5\, u(n)] = 2.5\, \frac{z}{z-1}$$

b) $x(n) = 2sin(\frac{\pi n}{4})$

*From table*

$$X(z) = Z[x(n)] = Z\left[2sin(\frac{\pi n}{4})\right] = 2\, \frac{z\, sin(\frac{\pi}{4})}{z^2 - 2z\, cos(\frac{\pi n}{4}) + 1}$$

b)      $x(n) = (0.8)^n\, u(n)$

$$X(z) = Z[x(n)] = Z[(0.8)^n u(n)] = \frac{z}{z - 0.8}$$


$x(n) = (0.8)^n\, sin(\frac{\pi n}{4})\, u(n)$

$X(z) = Z[x(n)] = Z\left[(0.8)^n sin(\frac{\pi n}{4})\, u(n)\right]$

$$= \frac{0.8\, z\, sin(\frac{\pi}{4})}{z^2 - 2*0.8\, z\, cos(\frac{\pi n}{4}) + 0.8^2}$$


a)      $x(n) = e^{-0.5n} cos\left(\frac{\pi n}{4}\right) u(n)$

$X(z) = Z[x(n)] = Z\left[e^{-0.5n} cos\left(\frac{\pi n}{4}\right) u(n)\right]$

$$= \frac{z[z - e^{-0.5}\, cos\left(\frac{\pi}{4}\right)]}{z^2 - 2*e^{-0.5}z\, cos(\frac{\pi n}{4}) + (e^{-0.5})^2}$$

## 8.3 Properties of the z-Transform

In this section, we study some important properties of the z-transform these properties widely used in driving the z- transform functions of difference equations and solving the system output responses of linear digital systems with constant system coefficients. For the following

$$Z\{f[n]\} = \sum_{n=0}^{n=\infty} f[n]z^{-n} = F(z)$$

(8.5)

$$Z\{g_n\} = \sum_{n=0}^{n=\infty} g_n z^{-n} = G(z)$$

(8.6)

- Linearity:
$$Z\{af_n + bg_n\} = aF(z) + bG(z).$$

(8.7)

   and ROC is $R_f \cap R_g$

which follows from the definition of z-transform.

- Time Shifting
If we have $f[n] \Leftrightarrow F(z)$ then $f[n - n_0] \Leftrightarrow z^{-n_0} F(z)$

(8.8)

The ROC of $Y(z)$ is the same as $F(z)$ except that there are possible pole additions or deletions at $z = 0$ or $z = \infty$.

Proof:
Let $y[n] = f[n - n_0]$ then

$$Y(z) = \sum_{n=-\infty}^{\infty} f[n - n_0]z^{-n}$$

(8.9)

Assume $k = n - n_0$ then $n = k + n_0$, substituting in the above equation we have:

$$Y(z) = \sum_{k=-\infty}^{\infty} f[k]z^{-k-n_0} = z^{-n_0} F[z]$$

(8.10)

- Multiplication by an Exponential Sequence

213

Let $y[n] = z_0^n f[n]$ then $Y(z) = X\left(\dfrac{z}{z_0}\right)$

$$(8.11)$$

The consequence is pole and zero locations are scaled by $z_0$. If the ROC of $FX(z)$ is $r_R < |z| < r_L$, then the ROC of $Y(z)$ is $r_R < |z/z_0| < r_L$, i.e., $|z_0|r_R < |z| < |z_0|rL$

Proof:

$$Y(z) = \sum_{n=-\infty}^{\infty} z_0^n x[n] z^{-n} = \sum_{n=-\infty}^{\infty} x[n] \left(\frac{z}{z_0}\right)^{-n} = X\left(\frac{z}{z_0}\right)$$

$$(8.12)$$

The consequence is pole and zero locations are scaled by $z_0$. If the ROC of $X(z)$ is $rR < |z| < rL$, then the ROC of $Y(z)$ is $rR < |z/z_0| < rL$, i.e., $|z_0|rR < |z| < |z_0|rL$

- Differentiation of $X(z)$

If we have $f[n] \Leftrightarrow F(z)$ then $nf[n] \xleftarrow{\ z\ } -z\dfrac{dF(z)}{z}$ and ROC $= R_f$ (8.13)

Proof:

$$F(z) = \sum_{n=-\infty}^{\infty} f[n] z^{-n}$$

$$-z\frac{dF(z)}{dz} = -z \sum_{n=-\infty}^{\infty} -n\, f[n] z^{-n-1} = \sum_{n=-\infty}^{\infty} -n\, f[n] z^{-n}$$

$$-z\frac{dF(z)}{dz} \xleftarrow{\ z\ } nf[n]$$

$$(8.14)$$

- **Conjugation of a Complex Sequence**

If we have $f[n] \Leftrightarrow F(z)$ then $f^*[n] \xleftarrow{\ z\ } F^*(z^*)$ and ROC $= R_f$

$$(8.15)$$

**Proof:**

Let $y[n] = f^*[n]$, then

$$Y(z) = \sum_{n=-\infty}^{\infty} f^*[n]z^{-n} = \left( \sum_{n=-\infty}^{\infty} f[n][z^*]^{-n} \right)^* = F^*(z^*)$$

(8.16)

- **Time Reversal**

If we have $f[n] \Leftrightarrow F(z)$ then $f^*[-n] \xleftarrow{\ z\ } F^*(1/z^*)$

(8.17)

Let $y[n] = f^*[-n]$, then

$$Y(z) = \sum_{n=-\infty}^{\infty} f^*[-n]z^{-n} = \left( \sum_{n=-\infty}^{\infty} f[-n][z^*]^{-n} \right)^* = \left( \sum_{k=-\infty}^{\infty} f[k](1/z^*)^{-k} \right)^* = F^*(1/z^*)$$

(8.18)

If the ROC of $F(z)$ is $r_R < |z| < r_L$, then the ROC of $Y(z)$ is

$$r_R < |1/z^*| < r_L \qquad i.e., \qquad \frac{1}{r_R} > |z| > \frac{1}{r_L}$$

When the time-reversal is without conjugation, it is easy to show

$$f[-n] \xleftarrow{\ z\ } F(1/z)$$

(8.19)

and ROC is $\dfrac{1}{r_R} > |z| > \dfrac{1}{r_L}$

A comprehensive summary of the z-transform properties shown in Table 8.2

**Table 8.2** Summary of z-transform properties

| No. | Property | Sequence | z-transform |
|-----|----------|----------|-------------|
| 1 | Addition | $x_1(n) + x_2(n)$ | $X_1(z) + X_2(z)$ |
| 2 | Constant multiple | $c\,x(n)$ | $c\,X(z)$ |
| 3 | Linearity | $ax_1(n) + bx_2(n)$ | $aX_1(z) + bX_2(z)$ |

215

| | | | |
|---|---|---|---|
| 4 | Delay unit step | u(n-m) | $\dfrac{Z^{1-m}}{Z-1}$ |
| 5 | Time delay shift | $x(n-m)\,U(n-m)$ | $Z^{-m}\,X(z)$ |
| 6 | Forward 1 tap | x(n+1) | $Z(X(z)-x(0))$ |
| 7 | Forward m tap | x(n+1) | $Z^m\left(X(z)-\displaystyle\sum_{i=0}^{m-1}x_i\,Z^{-i}\right)$ |
| 8 | Complex translation | $e^n\,x(n)$ | $X(ze^{-1})$ |
| 9 | Frequency scale | $b^n\,x(n)$ | $X(zb^{-1})$ |
| 10 | Differentiation | $n\,x(n)$ | $-Z\,X'(z)$ |
| 11 | Conjugation | $x^*(n)$ | $X^*(z^*)$ |
| 12 | Time reversal | $x(-n)$ | $X(z^{-1})$ |
| 13 | Integration | $\dfrac{1}{n}\,x(n)$ | $-\displaystyle\int\dfrac{X(z)}{Z}\,dz$ |
| 14 | Discrete-time convolution | $x_1(n)*x_2(n)$ | $X_1(z)\,X_2(z)$ |
| 15 | Initial time | $x(0)$ | $\displaystyle\lim_{n\to\infty}X(z)$ |
| 16 | Final value | $\displaystyle\lim_{n\to\infty}x(n)$ | $\displaystyle\lim_{n\to1}(Z-1)X(z)$ |

### Example 8.5

Find the z- transform of $x(n)= 3n + 2 \times 3^n$.

**Solution** From the linearity property

$Z\,[\,x(n)] = Z\,[3n + 2 \times 3^n] = 3Z\,[n] + 2\,Z\,[3^n]$

$$Z\{n\} = \frac{z}{(z-1)^2} \quad \text{and} \quad Z\{3^n\} = \frac{z}{(z-3)}$$

Therefore

$$Z\,[x(n)] = \frac{3z}{(z-1)^2} + \frac{2z}{(z-3)}$$

216

## Example 8.6

Find the z-transform of each of the following sequences:
(a) $x(n) = 2^n u(n) + 3(\frac{1}{2})^n u(n)$
(b) $x(n) = \cos(5n)u(n)$.

**Solution:**

(a) Because $x(n)$ is a sum of two sequences of the form $\alpha^n u(n)$, using the linearity property of the z-transform, and referring to Table 1, the z-transform pair

$$X(z) = \frac{1}{1-2z^{-1}} + \frac{3}{1-\frac{1}{2}z^{-1}} = \frac{4 - \frac{13}{2}z^{-1}}{\left(1-2z\right)\left(1-\frac{1}{2}z^{-1}\right)}$$

(b) For this sequence, we write

$x(n) = \cos(5n)\, u(n) = \frac{1}{2}(e^{j5n} + e^{-j5n})\, u(n)$
Therefore, the z-transform is

$$X(z) = \frac{1}{2}\frac{1}{1-e^{j5n}z^{-1}} + \frac{1}{2}\frac{1}{1-e^{-j5n}z^{-1}}$$

With a region of convergence $|z| > 1$. Combining the two terms, we have

$$X(z) = \frac{1 - z^{-1}\cos 5}{1 - 2z^{-1}\cos 5 + z^{-2}}$$

## Example 8.7
**Find the z-transform of the sequence given by**

$x(n) = u(n) - (0.85)^n u(n)$
**Solution**
$X(z) = Z[x(n)] = Z[u(n)] - Z[(0.85)^n u(n)]$

$$= \frac{z}{z-1} - \frac{z}{z-0.85}$$

## 8.4 Inverses z-transform

The z-transform of the sequence x(n)and the inverse z-transform of the function x(z) are defined as, respectively,

$$x(n) = \frac{1}{2\pi j} \oint X(z)z^{n-1}dz$$

(8.20)

**Where the circular symbol on the integral sign denotes a closed counter in the complex plane.**

The z-transform is a useful tool in linear systems analysis. However, just as important as techniques for finding the z-transform of a sequence are methods that may be used to invert the z-transform and recover the sequence $x(n)$ from $X(z)$. Three possible approaches are described below.

The inverse z-transform may be obtained by at least three methods:
1. Partial fraction expansion and look-up table
2. Power series expansion
3. Residue method

### 8.4.1 Partial fraction expansion and a look-up table
Now we are ready to deal with the inverse z-transform using the partial fraction
Expansion and look-up table. The general procedure is as follows:
1. Eliminate the negative powers of z for the z-transform function X (z)
2. Determine the rational function X(z)\z (assuming it is proper), and apply the partial fraction expanded function X(z)\z using the formula in Table 8.1.
3. Multiply the expanded function X(z)\z by z on both sides of the equation to obtain X(z).
4. Apply the inverse z- transform using table 1
The partial fraction format and the formula for calculating the constant are listed in table 3.
For z-transforms that are rational functions of z,

$$X(z) = \frac{\sum_{k=0}^{q} b(k)z^{-1}}{\sum_{k=0}^{p} a(k)z^{-1}} = C \frac{\prod_{k=0}^{q}(1 - \beta_k z^{-1})}{\prod_{k=1}^{p}(1 - \beta_k z^{-1})}$$

(8.21)

A simple and straightforward approach to find the inverse z-transform is to perform a partial fraction expansion of $X(z)$. Assuming that p > q and that all of the roots in the denominator are simple, $\alpha_i \neq \alpha_k$ for $i \neq k$, $X(z)$ expanded as follows:

218

$$X(z) = \sum_{k=1}^{p} \frac{A_k}{1 - \alpha_k z^{-1}}$$

$$(8.22)$$

for some constants $A_k$ for $k = 1, 2, \ldots p$. The coefficients $A_k$ may be found by multiplying both sides of Eq. (3) by $(1 - \alpha_k z^{-1})$ and setting $z = \alpha_k$. The result is

$$A_k = [(1 - \alpha_k z^{-1})X(z)]$$

$$(8.23)$$

If $p \leq q$, the partial fraction expansion must include a polynomial in $z^{-1}$ of order $(p\text{-}q)$. The coefficients of this polynomial found by long division (i.e., by dividing the numerator polynomial by the denominator). For multiple-order poles, the expansion must be modified. For example, if $X(z)$ has a second-order pole at $z = \alpha_k$, the expansion will include two terms,

$$\frac{B_1}{1 - \alpha_k z^{-1}} + \frac{B_2}{(1 - \alpha_k z^{-1})^2}$$

where $B_1$ and $B_2$ are given by

$$B_1 = \alpha_k \left[ \frac{d}{dz} (1 - \alpha_k z^{-1})^2 \, X(z) \right]$$

$$(8.24)$$

$$B_2 = [(1 - \alpha_k z^{-1})^2 \, X(z)]$$

$$(8.25)$$

### Example 8.8

Find the inverse of the following z-transform

$$X(z) = \frac{1}{(1 - z^{-1})(1 - 0.8z^{-1})}$$

**Solution:**
Multiplying the numerator and the denominator by $z^2$ we get

$$X(z) = \frac{1}{(1 - z^{-1})(1 - 0.8z^{-1})} \times \frac{z^2}{z^2} = \frac{z^2}{(z - 1)(z - 0.8)}$$

Dividing both sides by $z$, we have

$$\frac{X(z)}{z} = \frac{z}{(z - 1)(z - 0.8)}$$

We notice that the right-hand side of the above equation is a proper rational polynomial of $z$. Also, we see that the denominator of the

219

right-hand side has distinct poles, therefore, we right into partial fractions as,

$$\frac{X(z)}{z} = \frac{A}{(z-1)} + \frac{B}{(z-0.8)}$$

To find out the unknown constants $A$ and $B$, we use:

$$A = \left[(z-1) \times \frac{X(z)}{z}\right]_{z=1} = \left[(z-1) \times \frac{z}{(z-1)(z-0.8)}\right]_{z=1}$$

$$= \left[\frac{z}{(z-0.8)}\right]_{z=1} = \frac{1}{(1-0.8)} = 5$$

$$B = \left[(z-0.8) \times \frac{X(z)}{z}\right]_{z=0.8} = \left[(z-0.8) \times \frac{z}{(z-1)(z-0.8)}\right]_{z=0.8}$$

$$= \left[\frac{z}{(z-1)}\right]_{z=0.8} = \frac{0.8}{(0.8-1)} = -4$$

Substituting the values, we have,

$$\frac{X(z)}{z} = \frac{5}{(z-1)} + \frac{(-4)}{(z-0.8)}$$

Or it can be written as (by multiplying both sides by $z$)

$$X(z) = \frac{5z}{(z-1)} - \frac{4}{(z-0.8)}$$

Taking the inverse z-transform of both sides and using Table 1, we have

$$x(n) = Z^{-1}(X(z)) = 5Z^{-1}\left(\frac{z}{(z-1)}\right) - 4Z^{-1}\left(\frac{z}{(z-0.8)}\right)$$
$$= 5\,u(n) - 4(0.8)^n\,u(n)$$

**Example 8.9** Suppose that a sequence $x(n)$ has a z-transform

$$X(z) = \frac{4 - \frac{7}{4}z^{-1} + \frac{1}{4}z^{-2}}{1 - \frac{3}{4}z^{-1} + \frac{1}{8}z^{-2}}$$

**Solution:**

$$X(z) = \frac{4 - \frac{7}{4}z^{-1} + \frac{1}{4}z^{-2}}{1 - \frac{3}{4}z^{-1} + \frac{1}{8}z^{-2}} = \frac{4 - \frac{7}{4}z^{-1} + \frac{1}{4}z^{-2}}{(1 - \frac{1}{2}z^{-1})(1 - \frac{1}{4}z^{-1})}$$

With a region of convergence $|z| > \frac{1}{2}$. Because $p = q = 2$, and the two poles are simple, the partial fraction expansion has the form

$$X(z) = K_1 + \frac{K_2}{(1 - \frac{1}{2}z^{-1})} + \frac{K_3}{(1 - \frac{1}{4}z^{-1})}$$

The constant $K_1$ is found by long division, and equal to 2 .K2 and K3 are equal to 3 and -1 respectively.

Thus, the complete partial fraction expansion becomes

$$X(z) = 2 + \frac{3}{(1 - \frac{1}{2}z^{-1})} - \frac{1}{(1 - \frac{1}{4}z^{-1})}$$

Finally, because the region of convergence is the exterior of the circle $|z| > 1$, $x(n)$ is the right-sided sequence

$$x(n) = 2\delta(n) + 3(0.5)^n u(n) - (0.25)^n u(n)$$

## 8.4.2 Power Series

The z-transform is a power series expansion,

$$X(z) = \sum_{n=}^{\infty} x(n)z^{-n} = \cdots + x(-2)z^2 + x(-1)z^1 + x(0) + x(1)z^{-1}$$

$$+ x(2)z^{-2} + \cdots$$

(8.26)

where the sequence values $x(n)$ are the coefficients of $z^{-n}$ in the expansion, therefore, if we can find the power series expansion for $X(z)$, the sequence values $x(n)$ may be found by simply picking off the coefficients of $z^{-n}$.

### Example 8.10
Consider the z-transform

$$X(z) = \log(\frac{Z + c}{Z}) \qquad |Z| > |c|$$

### Solution:
The power series expansion of this function is

$$X(z) = \log(\frac{Z + c}{Z}) = \log(1 + c\, Z^{-1}) \qquad |Z| > |c|$$

$$= \sum_{n=1}^{\infty} \frac{1}{n}(-1)^{n+1} c^n\, Z^{-n} \qquad n > 0$$

Therefore, the sequence x(n) having this z-transform is

$$x(n) = \begin{cases} \frac{1}{n}(-1)^{n+1} c^n & n > 0 \\ \\ 0 & otherwise \end{cases}$$

**Exercise:**
Find the inverse of each of the following z-transforms:

i. $X(z) = 1 + 2(Z^2 + Z^{-2})$      $0 < |z| < \infty$

ii. $X(z) = \dfrac{1}{1-0.5Z^{-1}} + \dfrac{1}{1-0.2Z^{-1}}$      $0.5 < |z|$

iii. $X(z) = \dfrac{1}{1+2Z^{-1}+Z^{-2}}$      $|z| > 2$

iv. $X(z) = \dfrac{1}{(1-Z^{-1})(1-Z^{-2})}$      $|z| > 1$

## Example 8.11
find out the z-transform for the following sequences:
a.      $x(n) = 15\, u(n)$
b.      $x(n) = 10 \sin(0.25\pi n)\, u(n)$
c.      $x(n) = (0.5)^n\, u(n)$
d.      $x(n) = (0.5)^n \sin(0.25\pi n) u(n)$
e.      $x(n) = e^{-0.1n} \cos(0.25\pi n)\, u(n)$

**Solution:**

(a)      From Table 8.1, we get
$$X(z) = Z(x(n)) = Z(15\, u(n)) = \frac{15}{z-1}$$

(b)      From Table 8.1, we obtain
$$X(z) = Z(x(n)) = Z(10 \sin(0.25\pi n)\, u(n))$$
$$= \frac{10 \sin(0.25\pi)\, z}{z^2 - 2z\cos(0.25\pi n) + 1} = \frac{7.07\, z}{z^2 - 1.414z + 1}$$

(c)      From Table 8.1, we get
$$X(z) = Z(x(n)) = Z((0.5)^n\, u(n)) = \frac{z}{z-0.5}$$

(d)      From Table 8.1, we get
$$X(z) = Z(x(n)) = Z((0.5)^n \sin(0.25\pi n)u(n))$$
$$= \frac{0.5 \times \sin(0.25\pi)\, z}{z^2 - 2 \times 0.5 \times z \cos(0.25\pi n) + (0.5)^2}$$
$$= \frac{0.3536\, z}{z^2 - 1.4142z + 0.25}$$

(e)      From Table 8.1, we get

$$X(z) = Z(x(n)) = Z\left(e^{-0.1n} \cos(0.25\pi n)\, u(n)\right)$$
$$= \frac{z\left(z - e^{-0.1}\cos(0.25\pi)\right)}{z^2 - 2\,z\,e^{-0.1}\cos(0.25\pi) + (e^{-0.1})^2}$$
$$= \frac{z(z - 0.6397)}{z^2 - 1.279\,z + 0.8187}$$

## Example 8.12

Find the inverse of the following z-transform,
$$X(z) = \frac{1}{(1 - z^{-1})(1 - 0.5z^{-1})}$$

Eliminating the negative power of $z^2$ by multiply on the number or and determine by

$$X(z) = \frac{z^2}{z^2(1 - z^{-1})(1 - 0.5z^{-1})}$$

Dividing both sides by z leads to

$$\frac{X(z)}{z} = \frac{z}{(z - 1)(z - 0.5)}$$

$$= \frac{A}{(z - 1)} + \frac{B}{(z - 0.5)}$$

$$A = \lim_{z=1} \frac{z}{(z - 0.5)} = 2$$

$$B = \lim_{z=0.5} \frac{z}{(z - 1)} = 1$$

$$x(n) = Z^{-1}(X(z)) = Z^{-1}\left(\frac{z}{(z - 1)}\right) + Z^{-1}\left(\frac{z}{(z - 0.5)}\right)$$
$$= u(n) + (0.5)^n\, u(n)$$

## Example 8.13

Find the inverse of the following z-transform

i.
$$X(z) = 1 + \frac{z}{(z-1)} + \frac{z}{(z-0.8)}$$

ii.
$$X(z) = \frac{2z}{(z-1)^2} + \frac{5z}{(z-0.8)^2}$$

iii.
$$X(z) = \frac{z^{-6}}{z+1} + \frac{z^{-5}}{z-0.7} + z^{-4}$$

**Solution:**

Dividing both sides by $z$, we have

i.
$$X(z) = 1 + \frac{z}{(z-1)} + \frac{z}{(z-0.8)}$$

$$x(n) = Z^{-1}(X(z)) = Z^{-1}(1) + Z^{-1}\left(\frac{z}{(z-1)}\right) + Z^{-1}\left(\frac{z}{(z-0.8)}\right)$$
$$= \delta(n) + u(n) + (0.8)^n \, u(n)$$

ii.
$$X(z) = \frac{2z}{(z-1)^2} + \frac{5z}{(z-0.8)^2}$$

$$x(n) = Z^{-1}(X(z)) = Z^{-1}\left(\frac{2z}{(z-1)^2}\right) + Z^{-1}\left(\frac{5z}{(z-0.8)^2}\right)$$

$$= 2\,n\,u(n) + 5\,n\,(0.8)^n\,u(n)$$

$$= 2\,r(n) + 5\,(0.8)^n\,r(n)$$

iii.
$$X(z) = \frac{z^{-6}}{z+1} + \frac{z^{-5}}{z-0.7} + z^{-4}$$

224

$$x(n) = Z^{-1}(X(z)) = Z^{-1}\left(\frac{z^{-6}}{z+1}\right) + Z^{-1}\left(\frac{z^{-5}}{z-0.7}\right) + Z^{-1}(z^{-4})$$

$$x(n) = Z^{-1}\left(z^{-7}\frac{z}{z+1}\right) + Z^{-1}\left(z^{-6}\frac{z}{z-0.7}\right) + Z^{-1}(z^{-4})$$

$$x(n) = u(n-7) + (0.7)^n u(n-6) + \delta(n-4)$$

## Example 8.14

Determine the convolution of the following two sequences, using z-transform,

$$x_1(n) = 3\delta(n) + 2\delta(n-1)$$
$$x_2(n) = 2\delta(n) - \delta(n-1)$$

Solution

Taking the z-transform of the two sequences, we have
$$X_1(z) = Z(3\delta(n) + 2\delta(n-1)) = 3 + 2z^{-1}$$
$$X_2(z) = Z(2\delta(n) - \delta(n-1)) = 2 - z^{-1}$$
Using the z-transform property for convolution of two sequences, we have
$$X(z) = X_1(z)\, X_2(z) = (3 + 2z^{-1})(2 - z^{-1})$$
Therefore,
$$X(z) = 6 + z^{-1} - 2z^{-2}$$
Taking inverse z-transform of both sides (and using shift theorem), we have
$$x(n) = Z^{-1}(X(z)) = 6Z^{-1}(1) + Z^{-1}(z^{-1}) - 2Z^{-1}(z^{-2})$$
$$= 6\delta(n) + \delta(n-1) - 2\delta(n-2)$$

## Example 8.15
**Given a transfer function depicting a DSP system**
$$H(z) = \frac{z-1}{z+0.5}$$
**Determine**
(a)  The impulse response $h(n)$.
(b)  The step response $s(n)$.
(c)  The system response $y(n)$, if the input is given as $x(n) = (0.5)^n\, u(n)$.
Solution
Part (a): In this case $x(n) = \delta(n)$, thus $X(z) = 1$. As

225

$$H(z) = \frac{Y(z)}{X(z)}$$

Therefore, in this case, the z-transform of the output is equal to the transfer function:

$$H(z) = Y(z)$$

By taking the inverse z-transform of the transfer function, we can find out the unit impulse response $h(n)$, of the system. The transfer function is written as

$$\frac{H(z)}{z} = \frac{z - 1}{z(z + 0.5)}$$

This can further be written in the form of partial fractions as

$$\frac{H(z)}{z} = \frac{A}{z} + \frac{B}{(z + 0.5)}$$

where

$$A = \frac{z - 1}{(z + 0.5)}\bigg|_{z=0} = \frac{0 - 1}{(0 + 0.5)} = -2$$

$$B = \frac{z - 1}{z}\bigg|_{z=0.5} = \frac{0.5 - 1}{0.5} = -1$$

Thus we have

$$\frac{H(z)}{z} = \frac{-2}{z} + \frac{-1}{(z + 0.5)}$$

Or

$$H(z) = -2 - \frac{z}{(z - 0.5)}$$

Taking inverse z-transform of both sides (and using Table 5.1), we get

$$h(n) = -2\,\delta(n) - (0.5)^n\,u(n)$$

which is the required impulse response of the system.

Part (b): In this case $x(n) = u(n)$, thus $X(z) = \frac{z}{z-1}$. As

$$H(z) = \frac{Y(z)}{X(z)}$$

Therefore, in this case,

$$Y(z) = H(z)\,X(z) = \frac{(z - 1)}{(z + 0.5)}\frac{z}{(z - 1)}$$

It was written as

$$\frac{Y(z)}{z} = \frac{z - 1}{(z + 0.5)(z - 1)} = \frac{1}{(z + 0.5)}$$

Taking inverse z-transform of both sides, we get

226

$$y(n) = (-0.5)^n u(n)$$

which is the required step response of the system.

Part (c): In this case $x(n) = (0.25)^n u(n)$, $X(z) = \dfrac{z}{z-0.25}$

$$H(z) = \frac{Y(z)}{X(z)}$$

Therefore, in this case,

$$Y(z) = H(z)X(z) = \frac{(z-1)}{(z+0.5)} \cdot \frac{z}{(z-0.25)}$$

It was written as

$$\frac{Y(z)}{z} = \frac{z-1}{(z+0.5)(z-0.25)}$$

further can be written in the form of partial fractions as

$$\frac{Y(z)}{z} = \frac{A}{(z+0.5)} + \frac{B}{(z-0.25)}$$

where

$$A = \frac{z-1}{(z-0.25)}\bigg|_{z=-0.5} = \frac{-0.5-1}{(-0.5-0.25)} = 2$$

$$B = \frac{z-1}{(z+0.5)}\bigg|_{z=0.25} = \frac{0.25-1}{0.25+0.5} = -1$$

Thus we have

$$\frac{Y(z)}{z} = \frac{2}{(z+0.5)} + \frac{-1}{(z-0.25)}$$

Or

$$Y(z) = \frac{2z}{(z+0.5)} - \frac{z}{(z-0.25)}$$

Taking inverse z-transform of both sides, we get

$$y(n) = 2(-0.5)^n u(n) - (0.25)^n u(n)$$

**Exercises**

8.1 Given the sequence $x(n) = u(n-1)$. Find the z-transform of $x(n)$.

8.2 Given the sequence $x(n) = u(n+1)$. Find the z-transform of $x(n)$.

8.3 Considering the exponential sequence $x(n) = a^n u(n-1)$

Find the z-transform of the sequence $x(n)$.

8.4 Considering the exponential sequence $x(n) = a^n\ u(n + 1)$. Find the z-transform of the sequence x(n).

8.5 Find the z-transform for each of the following sequences $x(n)$.

f)   $x(n) = 2.5\ u(n - 1)$
g)   $x(n) = 102sin(\frac{\pi n}{4})$
h)   $x(n) = (0.8)^n\ u(n - 1)$
i)   $x(n) = (0.8)^{n-1}\ sin(\frac{\pi n}{4})\ u(n - 1)$
j)   $(n) = e^{-0.5n}cos\left(\frac{\pi n}{4}\right)u(n - 1)$

8.6 Find the z- transform of $x(n)= 3n + 2 \times 3^{n-1}$.

8.7 Find the z-transform of each of the following sequences:
(c)   $x(n) = 2^n u(n-1) + 3(½)^n u(n-1)$
(d)   $x(n) = cos(5n)u(n-1)$.

8.8 **Find the z-transform of the sequence given by**

$$x(n) = u(n + 1) - (0.85)^n u(n - 1)$$

8.9 **Find the z-transform of the sequence given by**
$$x_1(n) = \delta(n) - \delta(n - 1) + \delta(n - 2)$$
$$x_2(n) = 1.5\ \delta(n) + 2\ \delta(n - 1)$$
Also, find the convolution $x(n) = x_1(n) * x_2(n)$

8.10 Find the inverse of the following z-transform
$$X(z) = \frac{10}{(1 - 0.3\ z^{-1})(1 - 0.7z^{-1})}$$

8.11 Find the z-transform
$$X(z) = log(\frac{Z + c}{Z}) \qquad |Z| > |c|$$

8.12 Find the inverse of each of the following z-transforms:
■   $X(z) = 1 + 2(Z^3 + Z^{-3}) \qquad 0 < |z| < \infty$

- $X(z) = \dfrac{10}{1-0.5Z^{-1}} - \dfrac{10}{1-0.2Z^{-1}}$      $0.5 < |z|$

- $X(z) = \dfrac{10\, Z}{1+2Z^{-1}+Z^{-2}}$      $|z| > 2,$

- $X(z) = \dfrac{1}{Z(1-Z^{-1})(1-Z^{-2})}$      $|z| > 1$

8.13 Find the inverse $z$-transform of $X(z) = \cos z$.

8.14 **Given a transfer function depicting a DSP system**

$$H(z) = \frac{z-1}{z(z+0.5)}$$

**Determine**
(d)      **The impulse response $h(n)$.**
(e)      **The step response $s(n)$.**
(f)      **The system response $y(n)$, if the input is given as $x(n) = (0.8)^n\, u(n)$.**

8.15 **Given a transfer function depicting a DSP system**

$$H(z) = \frac{z(z-1)}{(z+0.5)}$$

**Determine**
(a)      **The impulse response $h(n)$.**
(b)      **The step response $s(n)$.**
(c)      **The system response $y(n)$, if the input is given as $x(n) = (0.75)^n\, u(n)$.**

**8.16** Find the z-transform for each of the following sequences (from the definition of the z-transform),
a.      $x(n) = 4\, u(n)$
b.      $x(n) = (0.7)^n\, u(n)$
c.      $x(n) = e^{-2n}\, u(n)$
d.      $x(n) = 2\,(0.8)^n \cos(0.2\pi n)\, u(n)$
e.      $x(n) = 2.5\, e^{-3n} \sin(0.2\pi n)\, u(n)$

**8.17** Using the properties of the z-transform, find the z-transform for each of the following sequences
a.      $x(n) = u(n-1) + (0.5)^n\, u(n)$
b.      $x(n) = e^{-3(n-5)} \cos\big(0.1\pi(n-5)\big)\, u(n-5)$,
where $u(n-5) = 1$ for $n \geq 5$ and $u(n-5) = 0$ for $n < 5$.

# Chapter 9

## Z-Transform Applications

Z- transform is used in many applications of mathematics and signal processing. The applications of z transform are: analyze the discrete linear system, finding frequency response Analysis of discrete signal, helps in system design and analysis and also checks the systems stability and analysis of digital filters. This chapter will focus on some these applications.

### 9.1 Evaluation of LTI System Response Using Z-Transform

Figure 9.1 shows an LTI System Where h(n) is the impulse response of the system, and H(z) is the transfer function, the input signal x(n) and the output response is y(n), the transfer function in z-domain given as

$$H(z) = H(e^{j\omega}) = \frac{Y(z)}{X(z)} \quad (9.1)$$

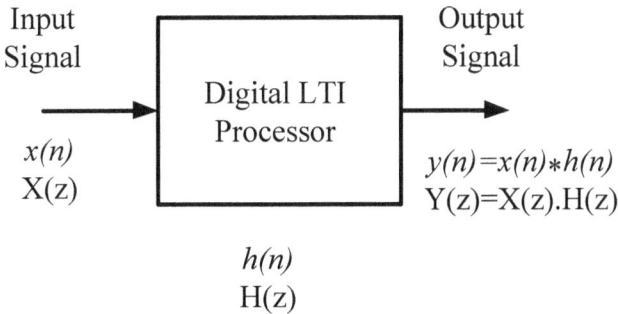

Input Signal

Output Signal

Digital LTI Processor

$x(n)$
$X(z)$

$y(n)=x(n)*h(n)$
$Y(z)=X(z).H(z)$

$h(n)$
$H(z)$

Figure 9.1 LTI processor system

### 9.2 Digital system implementation from its function

Since the z-transform is a linear transformation, the system implementation produce is similar to that in the time domain. The most convenient form for system synthesis is the z-transform of the general difference equation given by

$$Y(z) = \sum_{k=1}^{M} a_k z^{-k} Y(z) + \sum_{k=q}^{b} b_k z^{-k} X(z) \quad (9.2)$$

1-Gain: Figure 9.2 shows a Gain block, where k is the value of gain.

Figure 9.2 Gain block

2-Delay: Figure 9.3 shows a delay block.

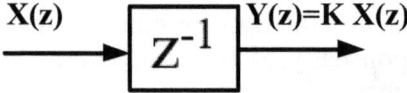

Figure 9.3 Delay block

3-Addition Figure 9.3 shows an addition block used to add two or more signals.

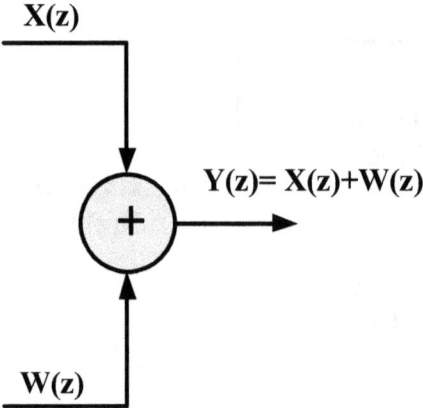

Figure 9.4 Addition block

Example 9.1
Find the impulse response and the transfer function of the following
system as shown in Figure

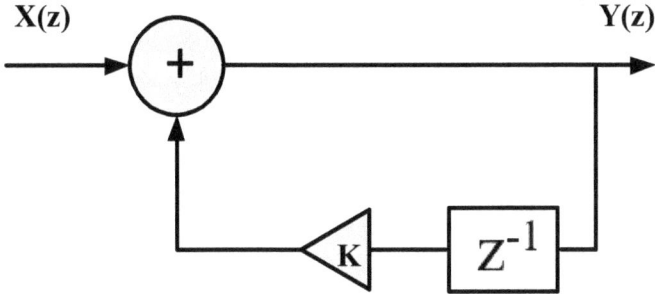

Figure 9.5 System for Example 9.1

Solution

$$y(n) = x(n) + K\,y(n-1)$$

$$Y(z) = X(z) + K\,Y(z)z^{-1}$$

For impulse response $x(n) = \delta(n)$
X(z)=1

$$Y(z) = 1 + K\,Y(z)z^{-1}$$

$$Y(z) = \frac{1}{1 - K\,z^{-1}}$$

Example 9.2

Implement 2nd order recursive filter for the sequence
$$y(n) = 2\,r\cos\omega_o\,y(n-1) - r^2\,y(n-2) + x(n) - r\cos\omega_o\,x(n-1)$$

Solution
Take z-transform for both sides to get

$$Y(z) = 2\,r\cos\omega_o\,Y(z)z^{-1} - r^2\,Y(z)z^{-2} + X(z) - r\cos\omega_o\,X(z)z^{-1}$$

So, the system implement as shown in Figure

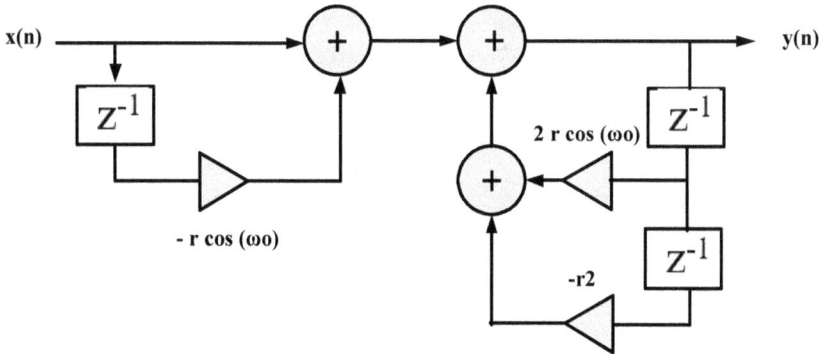

Figure 9.6 System for Example 9.2

**Given a second-order transfer function**
$$H(z) = \frac{2.5\,(1 - z^{-1})}{(1 + 1.3\,z^{-1} + 0.36\,z^{-2})}$$
**Perform the filter realizations and write the difference equations using the Cascade form realizations via the first-order sections**
**Solution**
**For the cascade realization, the transfer function is written in the product form. The given transfer function is**
$$H(z) = \frac{2.5(1 - z^{-2})}{1 + 1.3\,z^{-1} + 0.36\,z^{-2}}$$
**The numerator polynomial factorized as**
$$B(z) = 2.5(1 - z^{-2}) = 2.5\,(1 - z^{-1})(1 + z^{-1})$$
**The denominator polynomial factorized as**
$$\begin{aligned} A(z) &= 1 + 1.3\,z^{-1} + 0.36\,z^{-2} \\ &= 1 + 0.4\,z^{-1} + 0.9\,z^{-1} + 0.36\,z^{-2} \\ &= 1(1 + 0.4\,z^{-1}) + 0.9z^{-1}(1 + 0.4\,z^{-1}) \\ &= (1 + 0.4\,z^{-1})(1 + 0.9z^{-1}) \end{aligned}$$
**Therefore, the transfer function is written as**

$$H(z) = \frac{2.5\left(1 - z^{-1}\right)\left(1 + z^{-1}\right)}{(1 + 0.4\, z^{-1})(1 + 0.9z^{-1})}$$

Or

$$H(z) = \left(\frac{2.5 - 2.5\, z^{-1}}{1 + 0.4\, z^{-1}}\right)\left(\frac{1 + z^{-1}}{1 + 0.9z^{-1}}\right) = H_1(z) \cdot H_2(z)$$

Thus, in this case

$$H_1(z) = \frac{2.5 - 2.5\, z^{-1}}{1 + 0.4\, z^{-1}}$$

$$H_2(z) = \frac{1 + z^{-1}}{1 + 0.9z^{-1}}$$

Each one of $H_1(z)$ and $H_2(z)$ Can be realized in direct form I or direct form II. Overall, we get the cascaded realization with two sections. It should be noted that there could be other forms for $H_1(z)$ and $H_2(z)$, for example, we could have taken $H_1(z) = \frac{1+z^{-1}}{1+0.4\, z^{-1}}$, $H_2(z) = \frac{2.5-2.5\, z^{-1}}{1+0.9z^{-1}}$, to yield the same $H(z)$. Using the former $H_1(z)$ and $H_2(z)$, and using direct form II realizations for the two cascaded sections, we get the following difference equations:

Section 1: $\left(H_1(z) = \frac{2.5-2.5\, z^{-1}}{1+0.4\, z^{-1}}\right)$

$$w_1(n) = x(n) - 0.4\, w(n-1)$$
$$y_1(n) = 2.5\, w_1(n) - 2.5\, w_1(n-1)$$

Section 2: $\left(H_2(z) = \frac{1+z^{-1}}{1+0.9z^{-1}}\right)$

$$w_2(n) = y_1(n) - 0.9\, w_2(n-1)$$
$$y(n) = w_2(n) + w_2(n-1)$$

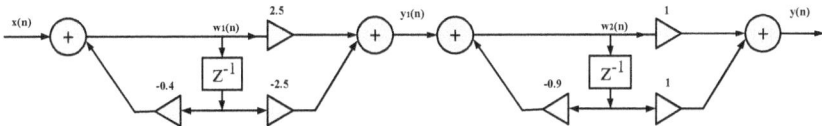

Figure 9.7 System for Example 9.3

**Example 9.4**

A relaxed (zero initial conditions) DSP system is described by the difference equation

235

$$y(n) + 0.1y(n-1) - 0.2y(n-2) = x(n) + x(n-1)$$

Determine the impulse response $y(n)$ due to the impulse sequence $x(n) = \delta(n)$.

**Solution:**

Taking z-transform of both sides of the given equation, we get

$$Z\big(y(n)\big) + 0.1Z\big(y(n-1)\big) - 0.2Z\big(y(n-2)\big)$$
$$= Z\big(x(n)\big) + Z(x(n-1)) \tag{1}$$

We have

$$Z\big(y(n)\big) = Y(z)$$
$$Z\big(x(n)\big) = X(z)$$

Using shift theorem, we have

$$Z\big(x(n-1)\big) = z^{-1}X(z)$$

Also, we can apply to sift theorem for $y$ in case of zero initial conditions, i.e.,

$$Z\big(y(n-1)\big) = z^{-1}Y(z)$$
$$Z\big(y(n-2)\big) = z^{-2}Y(z)$$

Putting these values in Equation (1), we have

$$Y(z) + 0.1z^{-1}Y(z) - 0.2z^{-2}Y(z) = X(z) + z^{-1}X(z)$$
$$\Rightarrow \qquad Y(z)\big(1 + 0.1z^{-1} - 0.2z^{-2}\big) = X(z)\big(1 + z^{-1}\big)$$

As $x(n) = \delta(n)$ therefore, (from Table 1), $X(z) = 1$. The above equation can now be written as

$$Y(z) = \frac{\big(1 + z^{-1}\big)}{\big(1 + 0.1z^{-1} - 0.2z^{-2}\big)}$$

Multiplying both the numerator and the denominator with $z^2$ we get

$$Y(z) = \frac{z(z+1)}{(z^2 + 0.1z - 0.2)}$$

The denominator factorized as

$$Y(z) = \frac{z(z+1)}{(z^2 + 0.5z - 0.4z - 0.2)}$$
$$= \frac{z(z+1)}{(z(z+0.5) - 0.4(z+0.5))}$$
$$= \frac{z(z+1)}{(z+0.5)(z-0.4)}$$
$$\Rightarrow \qquad \frac{Y(z)}{z} = \frac{(z+1)}{(z+0.5)(z-0.4)} \tag{2}$$

The right-hand side of the above equation is a proper rational polynomial, with the denominator polynomial having distinct poles. Therefore, it can be written into partial fractions as

236

$$\frac{Y(z)}{z} = \frac{A}{(z + 0.5)} + \frac{B}{(z - 0.4)} \tag{3}$$

To find out the unknown constants $A$ and $B$, we use:

$$A = \left[(z + 0.5) \times \frac{X(z)}{z}\right]_{z=-0.5}$$

$$= \left[(z + 0.5) \times \frac{(z + 1)}{(z + 0.5)(z - 0.4)}\right]_{z=-0.5}$$

$$= \left[\frac{(z + 1)}{(z - 0.4)}\right]_{z=-0.5} = \frac{(-0.5 + 1)}{(-0.5 - 0.4)} = \frac{0.5}{-0.9}$$

$$= -0.5556$$

$$B = \left[(z - 0.4) \times \frac{X(z)}{z}\right]_{z=0.4}$$

$$= \left[(z - 0.4) \times \frac{(z + 1)}{(z + 0.5)(z - 0.4)}\right]_{z=0.4}$$

$$= \left[\frac{(z + 1)}{(z + 0.5)}\right]_{z=0.4} = \frac{(0.4 + 1)}{(0.4 + 0.5)} = \frac{1.4}{0.9} = 1.5556$$

Equation (3) becomes:

$$\frac{Y(z)}{z} = \frac{-0.5556}{(z + 0.5)} + \frac{1.5556}{(z - 0.4)}$$

$$Y(z) = \frac{-0.5556\,z}{(z + 0.5)} + \frac{1.5556\,z}{(z - 0.4)}$$

Taking inverse z-transform of both sides

$$y(n) = Z^{-1}(Y(z)) = Z^{-1}\left(\frac{-0.5556\,z}{(z + 0.5)}\right) + Z^{-1}\left(\frac{1.5556\,z}{(z - 0.4)}\right)$$

$$= (-0.5556)Z^{-1}\left(\frac{z}{(z - (-0.5))}\right)$$

$$+ (1.5556)Z^{-1}\left(\frac{z}{(z - 0.4)}\right)$$

$$= (-0.5556)(-0.5)^n\, u(n)$$

$$+ (1.5556)(0.4)^n\, u(n)$$

Thus the output signal is

$$y(n) = (-0.5556)(-0.5)^n\, u(n) + (1.5556)(0.4)^n\, u(n)$$

**Example 9.5**

A relaxed (zero initial conditions) DSP system is described by the difference equation

$$y(n) = 0.4\,y(n - 1) + 0.32\,y(n - 2) + x(n) + 0.1\,x(n - 1)$$

Determine the impulse response $y(n)$ due to the impulse sequence $x(n) = \delta(n)$.

**Solution:**

$$y(n) - 0.4\,y(n-1) - 0.32\,y(n-2) = x(n) + 0.1\,x(n-1)$$

Taking z-transform of both sides of the given equation, we get

$$Z(y(n)) - 0.4Z(y(n-1)) - 0.32Z(y(n-2))$$
$$= Z(x(n)) + 0.1\,Z(x(n-1))$$

$$Y(z) - 0.4z^{-1}Y(z) - 0.32z^{-2}Y(z) = X(z) + 0.1z^{-1}X(z)$$
$$\Rightarrow \qquad Y(z)(1 - 0.4\,z^{-1} - 0.32\,z^{-2}) = X(z)(1 + 0.1\,z^{-1})$$

As $x(n) = \delta(n)$ therefore, (from Table 1), $X(z) = 1$. the above equation can now be written as

$$Y(z) = \frac{(1 + 0.1\,z^{-1})}{(1 - 0.4\,z^{-1} - 0.32\,z^{-2})}$$

Multiplying both the numerator and the denominator with $z^2$ we get

$$Y(z) = \frac{z(z + 0.1)}{(z^2 - 0.4\,z - 0.32)}$$

The denominator factorized as

$$Y(z) == \frac{z(z + 0.1)}{(z + 0.4)(z - 0.8)}$$

$$\Rightarrow \qquad \frac{Y(z)}{z} = \frac{(z + 0.1)}{(z + 0.4)(z - 0.8)}$$

The right-hand side of the above equation is a proper rational polynomial, with the denominator polynomial having distinct poles. Therefore, it can be written into partial fractions as

$$\frac{Y(z)}{z} = \frac{A}{(z + 0.4)} + \frac{B}{(z - 0.8)}$$

To find out the unknown constants $A$ and $B$, we use:

$$A = \left[(z + 0.4) \times \frac{X(z)}{z}\right]_{z=-0.4}$$

$$= \left[(z + 0.4) \times \frac{(z + 0.1)}{(z + 0.4)(z - 0.8)}\right]_{z=-0.4}$$

$$= \left[\frac{(z + 0.1)}{(z - 0.8)}\right]_{z=-0.4} = \frac{(-0.4 + 0.1)}{(-0.4 - 0.8)} = 0.25$$

$$B = \left[(z - 0.8) \times \frac{X(z)}{z}\right]_{z=0.8}$$

$$= \left[(z - 0.8) \times \frac{(z + 0.1)}{(z + 0.5)(z - 0.8)}\right]_{z=0.8}$$

$$= \left[\frac{(z + 0.1)}{(z + 0.4)}\right]_{z=0.8} = \frac{(0.4 + 0.1)}{(0.4 + 0.4)} = 0.625$$

Equation (3) becomes:

$$\frac{Y(z)}{z} = \frac{0.25}{(z + 0.4)} + \frac{0.625}{(z - 0.8)}$$

$$Y(z) = \frac{0.25\,z}{(z + 0.4)} + \frac{0.625\,z}{(z - 0.8)}$$

Taking inverse z-transform of both sides

$$y(n) = Z^{-1}(Y(z)) = Z^{-1}\left(\frac{0.25\,z}{(z + 0.4)}\right) + Z^{-1}\left(\frac{0.625\,z}{(z - 0.8)}\right)$$

$$= (0.25)Z^{-1}\left(\frac{z}{(z - (-0.4))}\right)$$

$$+ (0.625)Z^{-1}\left(\frac{z}{(z - 0.8)}\right)$$

$$= (0.25)(-0.4)^n\, u(n) + (0.625)(0.8)^n\, u(n)$$

Thus the output signal is

$$y(n) = (0.25)(-0.4)^n\, u(n) + (0.625)(0.8)^n\, u(n)$$

## 9.3 Pole Zero Diagrams for A Function in Z Domain

Z plane command computes and displays the pole-zero diagram of the Z function, as shown in Figure 9.8.

$$X(z) = \frac{0.8\,z^{-1} + z^{-1}}{1 - 2\,z^{-1} + 3\,z^{-1}}$$

**Poles:**

$$Z_{p1} = 1.0000 + 1.4142i$$
$$Z_{p2} = 1.0000 - 1.4142i$$

Zero:

$$Z_z = -1.2500$$

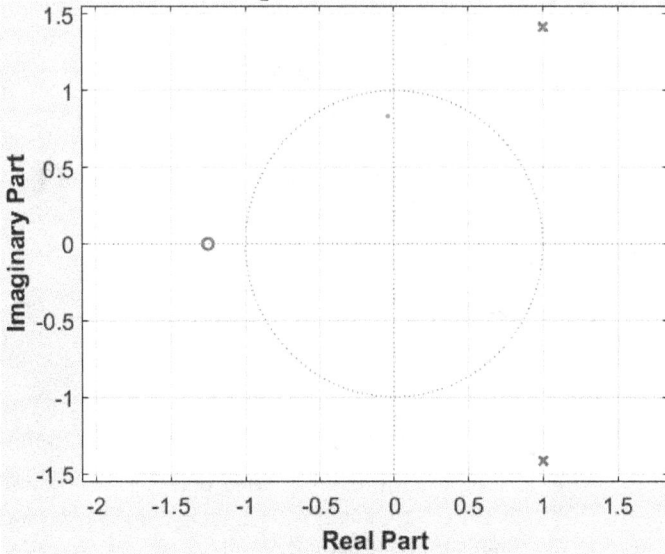

Figure 9.8 zero-pole diagram of the Z function

### 9.4 Frequency Response using z-transform:

The function computes and displays the frequency response of given Z- Transform
of the function, Figure 9.9 shows the Magnitude and phase response **using z-transform**

Fs= Sampling Frequency

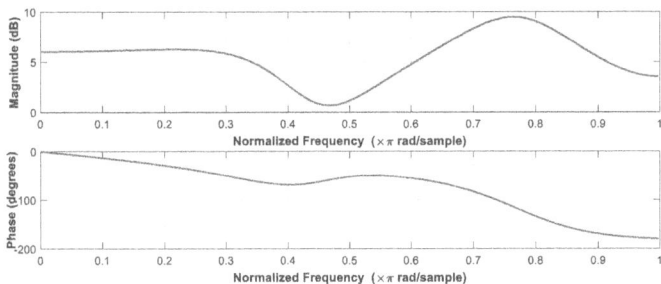

Figure 9.9 Magnitude and phase response **using z-transform**

### Example 9.6
Plot the magnitude and phase of the frequency response of the given digital filter

$$Y(n) = 1.5\ x(n) + 0.75y(n-1) - 0.9(y(n-2)$$

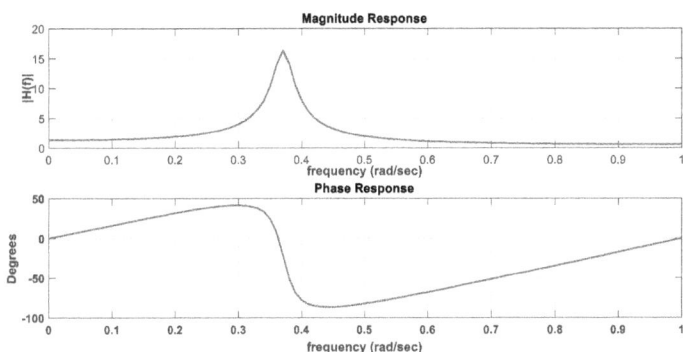

Figure 9.9 Magnitude and phase response **using z-transform for**

### Problems

9.1 Obtain the output for the following input x(n) and impulse response h(n) using z-transform

x(n)=10 u(n)

h(n)=5 u(n)

9.2 Find the impulse response and the transfer function of the following system as shown in Figure

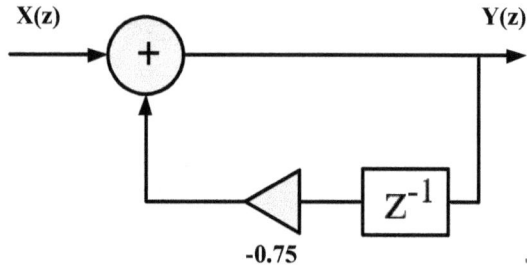

Figure 9.10 System for Problem 9.2

9.3 Find the impulse response and the transfer function of the following system as shown in Figure

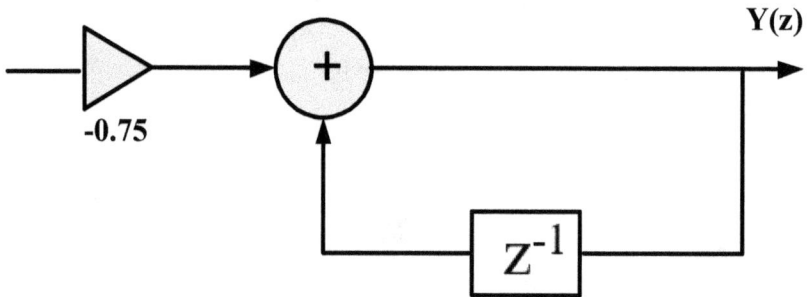

Figure 9.11 System for Problem 9.3

9.4 A signal x(n) begins at n=0 and has seven finite sample values [1 2 3 2 1 -1 1] if forms the input to an LTI processor whose impulse response h(n) begins at n=0 and has four finite sample values [1 1 1 1] convolute x(n) with h(n) to find output y(n) using z-transform.

**9.5** Given two sequences

$$x_1(n) = 5\,\delta(n) - 2\,\delta(n-2)$$
$$x_2(n) = 3\,\delta(n-3)$$

a. determine the z-transform of convolution of the two sequences using the convolution property of the z-transform,

$$X(z) = X_1(z)X_2(z)$$

242

b. determine convolution by the inverse z-transform from the result in part (a),

$$x(n) = Z^{-1}(X_1(z)X_2(z))$$

**9.6** Find the inverse z-transform for each of the following functions,

a. $X(z) = 4 - \dfrac{10z}{z-1} - \dfrac{z}{z+0.5}$

b. $X(z) = \dfrac{-5z}{(z-1)} + \dfrac{10z}{(z-1)^2} + \dfrac{2z}{(z-0.8)^2}$

c. $X(z) = \dfrac{z}{z^2+1.2z+1}$

d. $X(z) = \dfrac{4z^{-4}}{z-1} + \dfrac{z^{-1}}{(z-1)^2} + z^{-8} + \dfrac{z^{-5}}{z-0.5}$

**9.7** Using the partial fraction expansion method, find the inverse z-transform for each of the following functions,

a. $X(z) = \dfrac{1}{z^2-0.3z-0.04}$

b. $X(z) = \dfrac{2}{(z-0.2)(z+0.4)}$

c. $X(z) = \dfrac{z}{(z+0.2)(z^2-z+0.5)}$

d. $X(z) = \dfrac{z(z+0.5)}{(z-0.1)^2(z-0.6)}$

**9.8** A system is described by the difference equation

$$y(n) + 0.6y(n-1) = 4\,(0.8)^n u(n)$$

Determine the solution when the initial condition is $y(-1) = 2$.

**9.9** A system is described by the difference equation

$$y(n) - 1.5y(n-1) + 0.06\,y(n-2) = 2(0.4)^{n-1}u(n-1)$$

Determine the solution when the initial condition is $y(-1) = 1$ and $y(-2) = 1$.

**9.10** Given the following difference equation with the input-output relationship of a specific initially relaxed system (all initial conditions are zero),

$$y(n) - 0.7y(n-1) + 0.1\,y(n-2) = x(n) + 2x(n-1)$$

a. find the impulse response sequence $y(n)$ due to the impulse sequence $\delta(n)$.

b. find the output response of the system when the unit step function $u(n)$, is applied.

**9.11** Given the following difference equation with the input-output relationship of a specific initially relaxed DSP system (all initial conditions are zero),

$$y(n) - 1.4y(n-1) + 0.29\,y(n-2) = x(n) + 1.5x(n-1)$$

a.  find the impulse response sequence $y(n)$ due to the impulse sequence $\delta(n)$.

b.  find the output response of the system when the unit step function $u(n)$ is applied.

**9.12** Given the following difference equation,
$$y(n) = 0.5\,x(n) + 0.5\,x(n-1) + 0.5\,y(n-1)$$

a.  Find the transfer function $H(z)$;

b.  Determine the impulse response $y(n)$ if the input is $x(n) = 4\,\delta(n)$;

c.  Determine the step response $y(n)$ if the input is $x(n) = 10\,u(n)$.

**9.13** Given the following difference equation,
$$y(n) = x(n) - 0.5\,y(n-1) + 0.5\,y(n-2)$$

a.  Find the transfer function $H(z)$;

b.  Determine the impulse response $y(n)$ if the input is $x(n) = \delta(n)$;

c.  Determine the step response $y(n)$ if the input is $x(n) = u(n)$.

**9.14** Convert each of the following transfer functions into its difference equation

(a) $H(z) = \dfrac{z^2 - 0.25z}{z^2 + 1.1\,z + 0.18}$

(b) $H(z) = \dfrac{z^2 - 0.1\,z + 0.3}{z^3}$

**9.15** Given the following digital system with a sampling rate of 10000 Hz,
$$y(n) = 0.5\,x(n) + 0.5x(n-2)$$

c.  Determine the frequency response of the system;

d.  Calculate and plot the magnitude and phase frequency responses;

e.  Determine the filter type, based on the magnitude frequency response.

**9.16** Given the following digital system with a sampling rate of 10000 Hz,
$$y(n) = x(n) - 0.5y(n-2)$$

a. Determine the frequency response of the system;
b. Calculate and plot the magnitude and phase frequency responses;
c. Determine the filter type, based on the magnitude frequency response.

**9.17** Given the following difference equation for a digital system,
$$y(n) = x(n) - 2\cos(\alpha)\,x(n-1) + x(n-2) + 2\gamma\cos(\alpha) - \gamma^2$$
where $\gamma = 0.75$ and $\alpha = 30°$,
a. Find the transfer function $H(z)$;
b. Plot the poles and zeros on the z-plane with the unit circle;
c. Determine the stability of the system from the pole-zero plot;
d. Calculate the amplitude (magnitude) frequency response of $H(z)$;
e. Calculate the phase-frequency response of $H(z)$;

**9.18** Given the first order IIR system
$$H(z) = \frac{1 - 2\,z^{-1}}{1 - 0.5\,z^{-1}}$$
Realize $H(z)$ and develop the difference equations using the following forms:
    1. Direct-form I
    2-Direct-form II

**9.19** Given the filter
$$H(z) = \frac{1 - 0.9\,z^{-1} - 0.1\,z^{-2}}{1 - 0.3\,z^{-1} - 0.04\,z^{-2}}$$
Realize $H(z)$ and develop the difference equations using the following forms:
a. Direct-form I
b. Direct-form II
c. Cascade (series) form via the first-order sections
d. Parallel form via the first-order sections

**9.20** Given the filter
$$H(z) = \frac{1 + 2\,z^{-1} + z^{-2}}{1 - 0.5\,z^{-1} + 0.25\,z^{-2}}$$
plot:
a. Its magnitude frequency response;
b. Its phase frequency response.

9.21 Find the FFT for the input and output of the digital filter as shown in figure (1) for the input signal X (n) = [1 2 3 4].

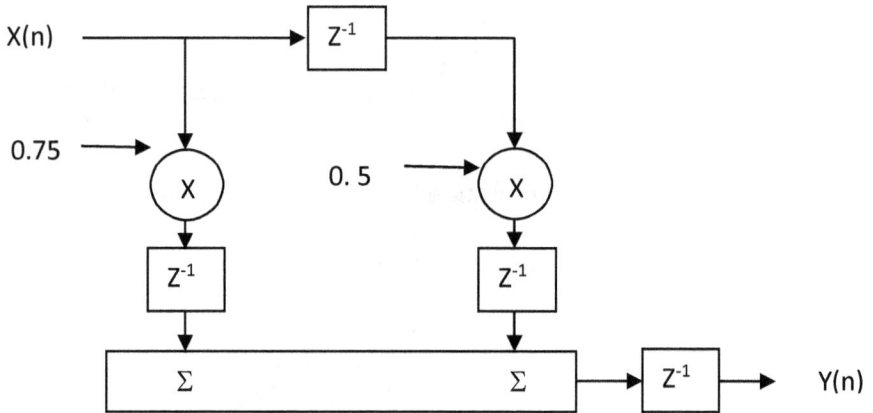

9.22 Express the following z-transform in factored form, plot its poles and zeros, and then determine its ROCs.

$$G(z) = \frac{4z^4 + 2z^3 + 4z^2 + 6z + 3}{3z^4 + 3z^3 - 1.5z^2 + z - 1.2}$$

9.23 Determine the partial fraction expansion of the z-transform G(z) given by

$$G(z) = \frac{3z^3}{2z^3 + 3z^2 - 4z - 1}$$

# Chapter 10

## Pole-Zero Stability

In DSP applications, the difference equation and transfer function are very important to study the characteristics of the system. The stability and frequency response can be examined based on the developed transfer function. This chapter will illustrate the concept of pole-zero stability, Stability determination-based z-transform, determination pole and zeros from difference equation and transfer function, and the stability rules.

### 10.1 Concept Poles and Zeros

A general causal digital filter has the difference equation:

$$y(n) = \sum_{i=0}^{N} a_i x(n-i) - \sum_{k=0}^{M} b_k x(n-k)$$

(10.1)

Which is of order max{ N, M }, and is recursive if any of the $b_j$ coefficients are non-zero. A second-order recursive digital filter, therefore, has the difference equation:

$$y[n] = a_0\,x[n] + a_1\,x[n-1] + a_2\,x[n-2] - b_1\,y[n-1] - b_2\,y[n-2]$$

A digital filter with a recursive linear difference equation can have an infinite impulse response. Remember that the frequency-response of a digital filter with impulse-response $\{h[n]\}$ is:

$$H\left(e^{j\Omega}\right) = \sum_{n=-\infty}^{\infty} h(n)e^{-j\Omega n}$$

(10.2)

### 10.1.1 Stability determination based z-transform

An LTI system can be described using z-transform as a ratio is

$$H(z) = \frac{N(z)}{D(z)} = \frac{k(z-z_1)(z-z_2)(z-z_3)\ldots}{(z-p_1)(z-p_2)(z-p_3)\ldots}$$

(10.3)

Where k is the system gain and The constants z1,z2,z3,....are called zeros pf X(z) because they are values of (z) for which H(z) is zero.Conversely p1,p2,p3,...are called poles of H(z). The poles and zeros are either real or complex conjugate numbers.
The digital system is stable, if and only all poles of the system lie inside the unit circle in the z plane

## 10.1.2 The z-transform:

Consider the response of a causal stable LTI digital filter to the special sequence $\{z^n\}$ where z is a complex. If $\{h[n]\}$ is the impulse-response, by discrete-time convolution, the output is a sequence $\{y[n]\}$ where

$$y(n) - \sum_{k=-\infty}^{\infty} h(n)z^{n-k} = z^n \sum_{k=-\infty}^{\infty} h(n)z^{-k}$$

$$= z^n H(z)$$

(10.4)

The expression obtained for H(z) is the 'z-transform' of the impulse response. H(z) is a complex number when evaluated for a given complex value of z.

It indicated that H(z) must be finite for a stable causal system when evaluated for a complex number z with modulus greater than or equal to one.

Since $H(z) = \sum_{n=-\infty}^{\infty} h[n]z^{-n}$ and the frequency - response: $H(e^{i\Omega}) = \sum_{n=-\infty}^{\infty} h[n]z^{-j\Omega n}$

It is clear that replacing z by e $^{j\Omega}$ in H(z) gives H(e$^{j\Omega}$).

## 10.1.3 The 'z-plane'

Conveniently, it is represented as complex numbers on an 'Argand diagram' as illustrated in **Figure 10.1**. The main reason for doing this is that the modulus of the difference between two complex numbers a+jb and c+jd say, i.e., | (a+jb) - (c+jd) |is represented graphically as the length of the line between the two complex numbers as plotted on the Argand diagram.

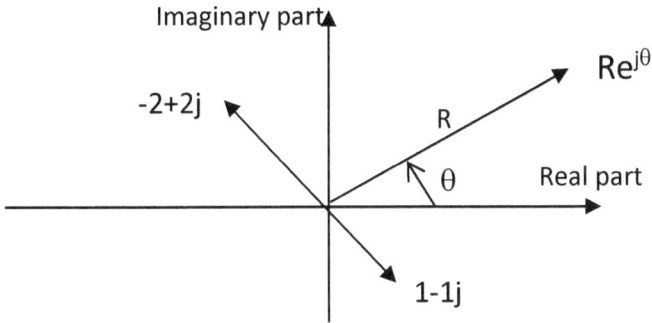

**Figure 10.1** complex number phasor diagram

If one of these complex numbers, c +jd say is zero, i.e., 0+j0, then the modulus of the other number |a+jb| is the distance of a+jb from the origin 0+j0 on the Argand diagram.

Of course, any complex number like a+jb can be converted to polar form $Re^{j\theta}$ where R= |a+jb| and $\theta$= $\tan^{-1}$(b/a). Plotting a complex number expressed as $Re^{j\theta}$ on an Argand diagram is illustrated above. We draw an arrow of length R starting from the origin and set it at an angle $\theta$ from the 'real part' axis (measured anti-clockwise). $Re^{j\theta}$ is then at the tip of the arrow. The illustration above, $\theta$ is about $\pi/4$ or 45 degrees. If R=1, $Re^{j\theta} = e^{j\theta}$ and the Argand diagram would be a point at a distance from the origin. We plot $e^{j\theta}$ for values $\theta$ in the range 0 to $2\pi$ 360° produces points, all of which lie on a 'unit circle,' i.e., a circle of radius 1, with center at the origin.

Where the complex numbers plotted on an Argand diagram are values of z for which we are interested in H(z), the diagram is referred to as 'the z-plane'. Points with z = $e^{j\Omega}$ lie on a unit circle, as shown in **Figure 10.2**. Remember that $|e^{j\Omega}|$ = $|\cos(\Omega) + j\sin(\Omega)|$ = $\sqrt{[\cos^2(\Omega) + \sin^2(\Omega)]}$ = 1. Therefore, evaluating the frequency-response $H(e^{j\Omega})$ for $\Omega$ in the range 0 to $\pi$ is equivalent to evaluating H(z) for z =$e^{j\Omega}$ which goes round the upper part of the unit circle as $\Omega$ goes from 0 to $\pi$.

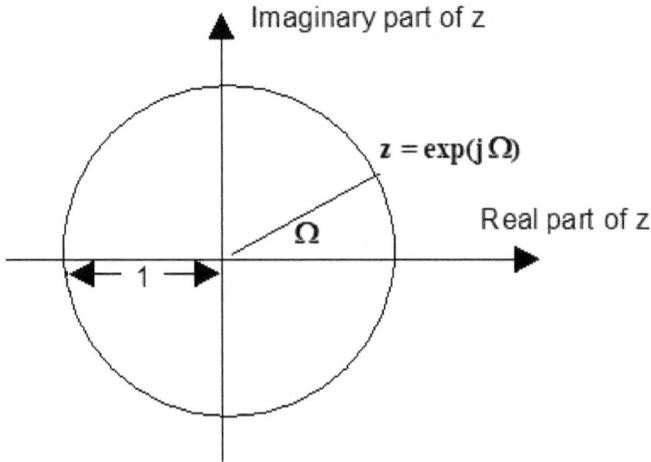

**Figure 10.2** z-plane diagram

## 10.2 Difference Equation and Transfer Function

Let The general difference equation given by

$$y(n) = b_0x(n) + b_0x(n-1) + \cdots + b_Mx(n-M) - a_1y(n-1) + \cdots - a_Ny(n-N)$$

$$(10.5)$$

The supposition that all initial conditions of this system are zero, and the X(z) and Y(z) representing the z-transforms of the sequences x(n) and y(n), respectively, taking the z-transform of Equation (9.1) yields

$$Y(z) = b_0X(z) + b_0X(z)z^{-1} + \cdots + b_MX(z)z^{-M} - a_1Y(z)z^{-1} + \cdots - a_NY(z)z^{-N}$$

$$(10.6)$$

The z-transfer function H(z) is defined as the ratio of z-transform of the output Y(z) to the z-transform of the input X(z).To get the transfer function Rearranging Equation (9.2)

$$H(z) = \frac{Y(z)}{X(z)} = \frac{b_0 + b_1z^{-1} + \cdots + b_Mz^{-M}}{1 + a_1z^{-1} + \cdots + a_Nz^{-N}} = \frac{A(z)}{B(z)}$$

where H(z) is defined as the transfer function with its numerator and denominator polynomials defined as

$$A(z) = b_0 + b_1 z^{-1} + \cdots + b_M z^{-M} \qquad (10.8)$$

$$B(z) = 1 + a_1 z^{-1} + \cdots + a_N z^{-N} \qquad (10.9)$$

The z-transfer function and represent the digital filter in the z-domain, as shown in **Figure 10.3**

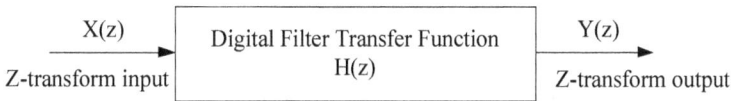

**Figure 10.3** z-transfer function and represent as the digital filter

Example 10.1

The following difference equation describes a DSP system
$$y(n) = x(n) + x(n-1) - 1.2x(n-2) - 2y(n-1) - 0.8y(n-2)$$
Find the transfer function H(z), the Numerator polynomial A(z), and the denominator polynomial equation B(z).
Solution:
Taking the z-transform on both sides of the previous difference equation, we achieve

Moving the last two terms to the left side of the difference equation and factoring Y(z)

On the left side X(z) on the right side, we obtain

$$y(n) = x(n) + x(n-1) - 1.2x(n-2) - 2y(n-1) - 0.8y(n-2)$$
$$Y(z) = X(z) + X(z)Z^{-1} - 1.2X(z)Z^{-2} - 2Y(z)Z^{-1} - 0.8Y(z)Z^{-2}$$

$$Y(z) + 2Y(z)Z^{-1} + 0.8Y(z)Z^{-2} = X(z) + X(z)Z^{-1} - 1.2X(z)Z^{-2}$$

$$[1 + 2Z^{-1} + 0.8\,Z^{-2}]Y(z) = [1 + Z^{-1} - 1.2Z^{-2}]\,X(z)$$

$$\frac{Y(z)}{X(z)} = \frac{1 + Z^{-1} - 1.2Z^{-2}}{1 + 2Z^{-1} + 0.8\,Z^{-2}}$$

$$\frac{Y(z)}{X(z)} = \frac{Z^2 + Z - 1.2}{Z^2 + 2Z + 0.8}$$

The Numerator polynomial $A(z) = Z^2 + Z - 1.2$
The denominator polynomial $B(z) = Z^2 + 2Z + 0.8$

Example 10.2

The following difference equation describes a system
$$y(n) = x(n) - 0.35x(n - 1) + 1.2x(n - 2))$$
Find the transfer function H(z), the Numerator polynomial A(z), and the denominator polynomial equation B(z).
Solution
Taking the z-transform on both sides of the previous difference equation, we achieve
$$y(n) = x(n) - 0.35x(n - 1) + 1.2x(n - 2))$$
$$Y(z) = X(z) - 0.35X(z)Z^{-1} + 1.2X(z)Z^{-2}$$

$$Y(z) = [1 - 0.35Z^{-1} + 1.2Z^{-2}]\,X(z)$$

$$\frac{Y(z)}{X(z)} = [1 - 0.35Z^{-1} + 1.2Z^{-2}]$$

The Numerator polynomial $A(z) = 1 - 0.35Z^{-1} + 1.2Z^{-2}$
The denominator polynomial $B(z) = 1$

## 10.3 BIBO stability

A system is said to be bounded input-bounded output stable (BIBO stable or just stable) if the output signal is bounded for all input signals that are bounded.

Consider a discrete-time system with input $x$ and output $y$. The input is bounded if there is a real number $M < \infty$ $|x(k)| \le M$ for all k.

An output is bounded if there is a real $N < \infty$ $|y(k)| \le N$ for n.

The system is stable if there is some bound N on the output for any input bounded by M.

Theorem:
A discrete-time LTI system is stable if its impulse response is summable.

Proof of if:

Consider a discrete-time LTI system with impulse response h. The convolution sum gives the output y corresponding to the input x,

$$\forall n \in Integers, \; y(n) = \sum_{m=-\infty}^{\infty} h(m)x(n-m) \qquad (10.10)$$

Suppose that the input bounded with bound M. Then, applying the triangle inequality, we see that

$$|y(n)| \le \sum_{m=-\infty}^{\infty} |h(m)||x(n-m)| \le M \sum_{m=-\infty}^{\infty} |h(m)| \qquad (10.11)$$

Thus, if the impulse response is summable, then the output is bounded with bound

$$N = M \sum_{m=-\infty}^{\infty} |h(m)|$$

$$(10.12)$$

Proof of only if:

To show that the system is not stable, we need to find one bounded input for which the output either does not exist or is not bounded. Such information is given by

$\forall n \in Integers,$

$$x(n) = \frac{h(-n)}{|h(-n)|}, h(n) \neq 0$$

$$= 0, h(n) = 0$$

(10.13)

The input is bounded, with bound M=1. Plugging this input to the convolution sum (1) and evaluating at n=0, we get

$$y(0) = \sum_{m=-\infty}^{\infty} h(m)x(-m) = \sum_{m=-\infty}^{\infty} \frac{(h(m))^2}{|h(m)|} = \sum_{m=-\infty}^{\infty} |h(m)| \qquad (10.14)$$

But since the assumption that the impulse response is not summable, y(0) does not exist or is finite, the system is not stable.

### 10.4 The z-Plane Pole-Zero Plot and Stability

A handy tool to analyze digital systems is the z-plane pole-zero plot. This graphical technique allows us to investigate the characteristics of the digital system shown in Figure 10.3, including system stability. In general, a digital transfer function can be written in the pole-zero form, and we can plot the poles and zeros on the z-plane. The z-plane is depicted in **Figure 10.4** and has the following features:

1. The horizontal axis is the real part of the variable z, and the vertical axis represents the imaginary part of the variable z.
2. The z-plane is divided into two parts by a unit circle.
3. Each pole is marked on the z-plane using the cross symbol ⌐, while each zero is plotted using the small circle symbol.

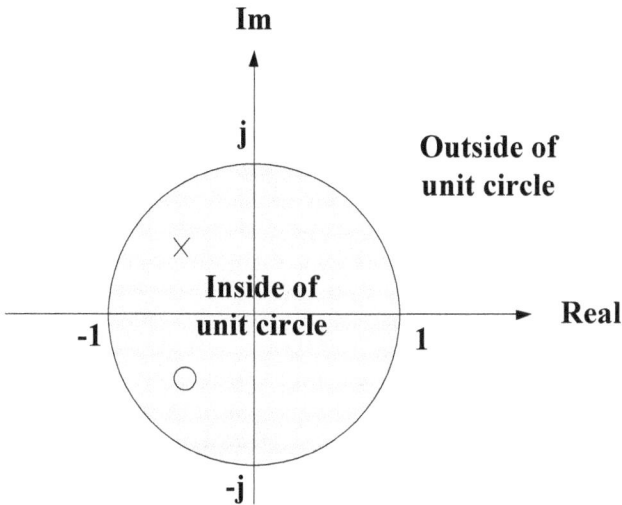

**Figure 10.4** z-Plane Pole-Zero Plot

## 10.5 Stability rules

Like the analog system, the digital system requires that all poles plotted on the z-plane be inside the unit circle. We summarize the rules for determining the stability of a DSP system as follows:

1. If the outermost pole(s) of the z-transfer function H(z) describing the DSP system is(are) inside the unit circle in the z-plane pole-zero plot, then the system is stable.
2. If the outermost pole(s) of the z-transfer function H(z) is(are) outside the unit circle in the z-plane pole-zero plot, the system is unstable.
3. If the outermost pole(s) is(are) first-order pole(s) of the z-transfer function H(z) and on the unit circle in the z-plane pole-zero plot, then the system is marginally stable.
4. If the outermost pole(s) is(are) multiple-order pole(s) of the z-transfer function H(z) and on the unit circle in the z-plane pole-zero plot, then the system is unstable.
5. The zeros do not affect system stability.

Notice that the following facts apply to a stable system (bounded-in/bounded-out
[BIBO] stability discussed in chapter 2):

1. If the input to the system is bounded, then the system's output will also be bounded, or the system's impulse response will go to zero in a finite number of steps.
2. An unstable system is one in which the system's output will grow without bound due to any bounded input, initial condition, or noise, or its impulse response will grow without bound.
3. The impulse response of a marginally stable system stays at a constant level or oscillates between two finite values.

Example 10.3

The example illustrating the  rules of stability
When the input is impulse sequence $x(n) = \delta(n)$ as in Figure 10.5 a and the difference output equation given as   $y(n) = x(n) + 0.75\, y(n-1)$, the transfer function is given as

$$H(z) = \frac{z}{z - 0.75} = Y(z)$$

And the output response can be written as  $y(n) = (0.75)^n\, u(n)$ as in Figure 10.5 b
The z plain given in Figure 10.5 c shows the system's stability.

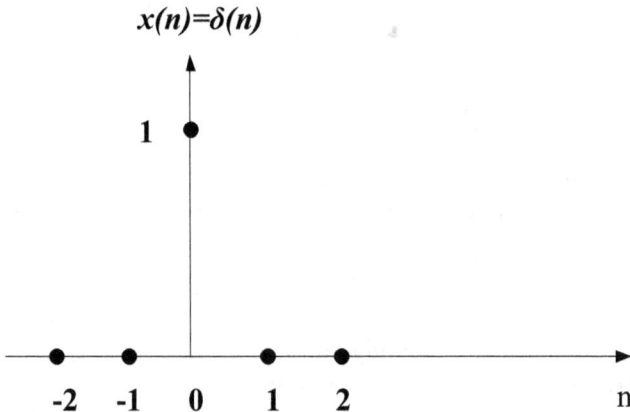

$x(n)=\delta(n)$

(a) Input signal

(b) Output sequence

(c) Z-plane representation

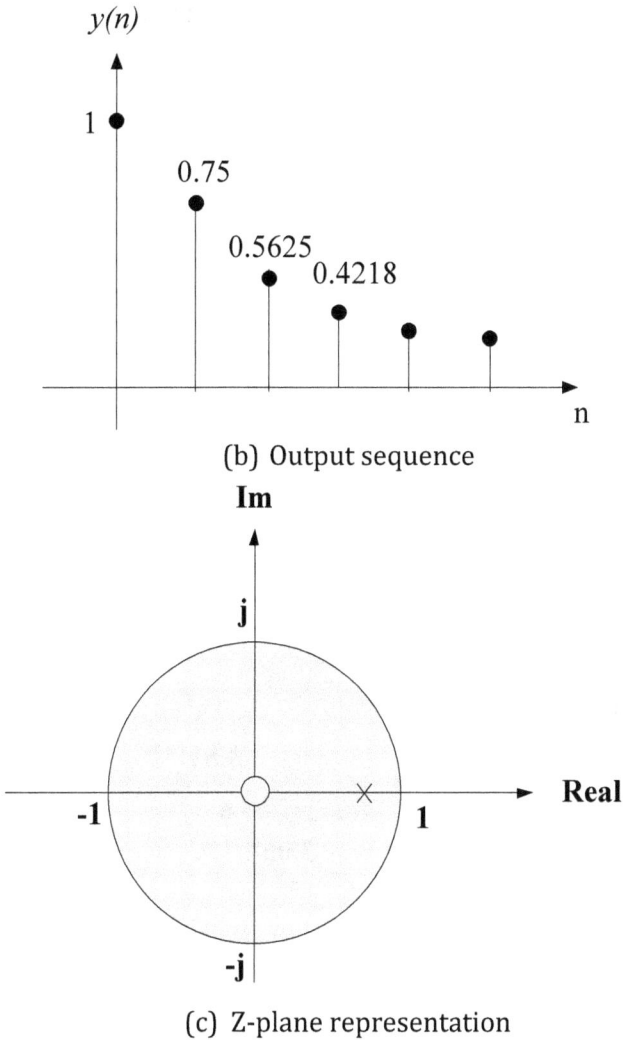

**Figure 10.5** Response of a stable system

When the input is impulse sequence $x(n) = \delta(n)$ as in Figure 10.6 a and the difference output equation given as $y(n) = x(n) + 1.25\, y(n-1)$, the transfer function is given as

$$H(z) = \frac{z}{z - 1.25} = Y(z)$$

And the output response can be written as $y(n) = (1.25)^n \, u(n)$ as in Figure 10.6 b

The z plain given in Figure 10.6 c shows the system's stability.

$$x(n)=\delta(n)$$

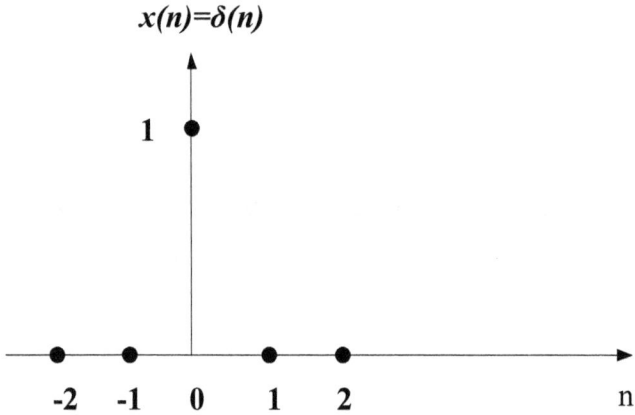

(a) Input signal

$$y(n)$$

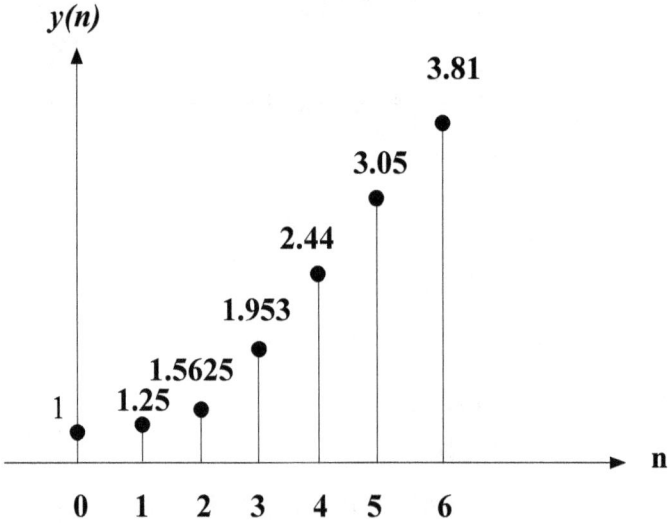

(b) Output sequence

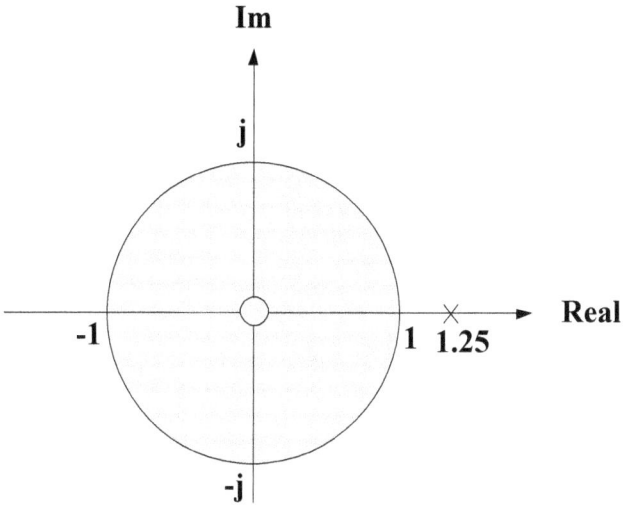

(c) Z-plane representation

**Figure 10.6** Response of the unstable system

When the input is impulse sequence $x(n) = \delta(n)$ as in Figure 10. and the difference output equation given as $y(n) = x(n) + y(n-1)$, the transfer function is given as

$$H(z) = \frac{z}{z-1} = Y(z)$$

And the output response can be written as $y(n) = u(n)$ as in Figure 10.
The z plane is given in Figure 10. And show the stability of the system.

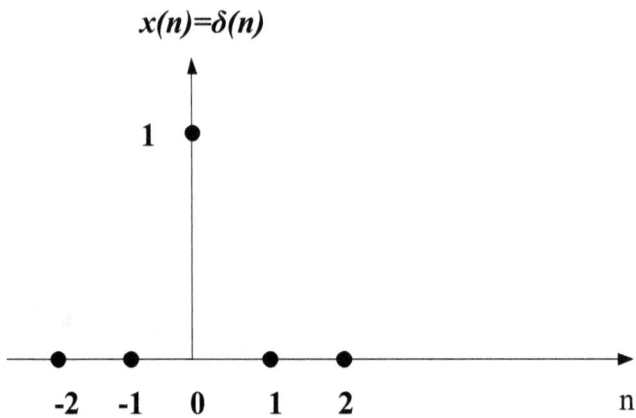

(a) Input signal

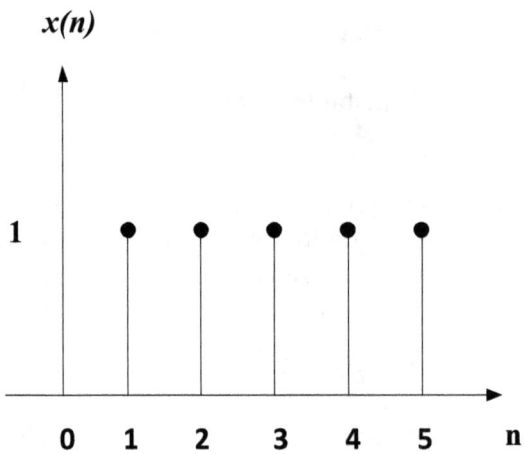

(b) Output sequence

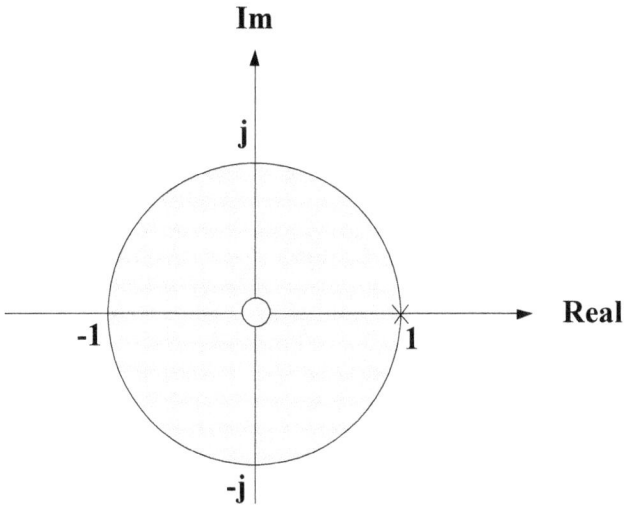

(c) Z-plane representation

**Figure 10.7** Response of the critical stable system

Example 10.4
Given the following transfer function

$$H(z) = \frac{z^{-1} - 0.6\,z^{-2}}{1 + 1.2z^{-1} + 0.55z^{-2}}$$

Convert it into its pole-zero form.
Solution
We first multiply the numerator and denominator by $z^2$ to obtain the transfer function whose numerator and the denominator polynomials have the positive powers of $z$, as follows

$$H(z) = \frac{z^2(z^{-1} - 0.6\,z^{-2})}{z^2(1 + 1.2z^{-1} + 0.55z^{-2})}$$

$$H(z) = \frac{z - 0.6}{z^2 + 1.2z + 0.55}$$

The zero of H(z): $z - 0.6 = 0 \rightarrow z = 0.6$
The poles of H(z): $z^2 + 1.2z + 0.55 = 0 \rightarrow z = -0.6 \pm j0.4359$

From Figure 10.8, it has seen the poles inside the unit circle, so the system is stable

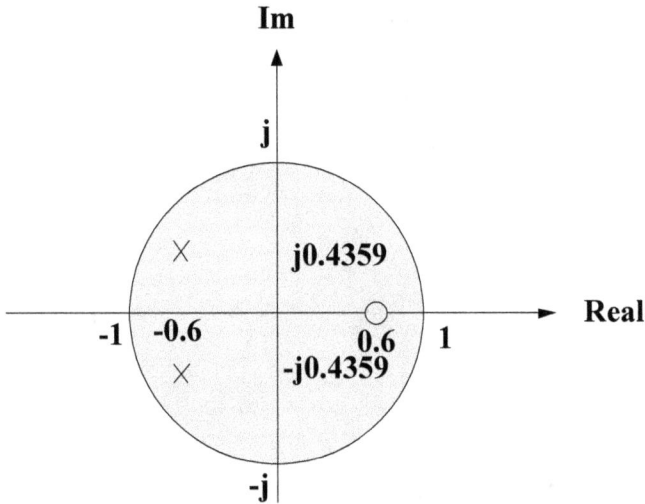

**Figure 10.8** Z-plane representation **of** Example 10.4

Example 10.5
Given the following transfer function

$$H(z) = \frac{z(1 - z^{-2})}{(1 + 1.3z^{-1} + 0.36z^{-2})}$$

Convert it into its pole-zero form.

Solution

We first multiply the numerator and denominator by $z^2$ to obtain the transfer function whose numerator and the denominator polynomials have the positive powers of $z$, as follows

$$H(z) = \frac{(1 - z^{-2})\, z^2}{(1 + 1.3z^{-1} + 0.36z^{-2})\, z^2} = \frac{z^2 - 1}{z^2 + 1.3z + 0.36}$$

Putting the numerator polynomial equal to zero and then finding the roots gives us the zeros of the transfer function,

$$z^2 - 1 = 0$$
$$(z - 1)(z + 1) = 0$$

Therefore, we get $z_1 = 1$ and $z_2 = -1$ as the roots.

Now, setting the denominator polynomial equal to zero and finding the roots gives us the poles of the transfer function,

$$z^2 + 1.3z + 0.36 = 0$$

$$z = \frac{-1.3 \pm \sqrt{(1.3)^2 - 4(1)(0.36)}}{2(1)} = \frac{-1.3 \pm \sqrt{1.69 - 1.44}}{2}$$

$$= \frac{-1.3 \pm \sqrt{0.25}}{2} = \frac{-1.3 \pm 0.5}{2} = -0.4, -0.9$$

Therefore, the poles are $p_1 = -0.4$ and $p_{21} = -0.9$. The transfer function can now be written in the pole-zero form as

$$H(z) = \frac{(z-1)(z+1)}{(z+0.4)(z+0.9)}$$

Example 10.6

The following transfer functions describe digital systems

(a) $H(z) = \frac{(z-0.5)}{(z-0.5)(z^2 + z + 0.5)}$

(b) $H(z) = \frac{(z^2 + 0.25)}{(z-0.5)(z^2 + 4z + 4.5)}$

(c) $H(z) = \frac{(z+0.25)}{(z-0.25)(z^2 +1.5 z + 1)}$

(d) $H(z) = \frac{(z^2 + z + 0.25)}{(z-1)^2(z+1)(z-0.6)}$

For each, sketch the z-plane pole-zero plot and determine the stability status for the digital system.

Solution

a- Putting the numerator polynomial equal to zero and then finding the roots,

$$z - 0.5 = 0$$

Therefore, we get $z_1 = 0.5$ as the root.

Now, setting the denominator polynomial equal to zero and finding the roots gives us the poles of the transfer function,

$$(z - 0.5)(z^2 + z + 0.5) = 0$$

This leads to

$$z - 0.5 = 0$$

and

$$z = \frac{-1 \pm \sqrt{(1)^2 - 4(1)(0.5)}}{2(1)} = \frac{-1 \pm \sqrt{1 - 2}}{2} = \frac{-1 \pm \sqrt{-1}}{2}$$

$$= \frac{-1.0 \pm j}{2} = -0.5 \pm j0.5$$

Therefore, the poles are $p_1 = 0.5$, $p_2 = -0.5 + j0.5$ and $p_3 = -0.5 - j0.5$. The magnitudes of these poles are

$$|p_1| = 0.5$$

$$|p_2| = |-0.5 + j0.5| = \sqrt{(-0.5)^2 + (0.5)^2} = 0.707$$

$$|p_3| = |-0.5 - j0.5| = \sqrt{(-0.5)^2 + (-0.5)^2} = 0.707$$

It can be noticed that the magnitudes of all the poles are less than 1, so they are inside the unit circle in the z-plane pole-zero plot. Therefore, the system is stable. It is shown in the following figure.

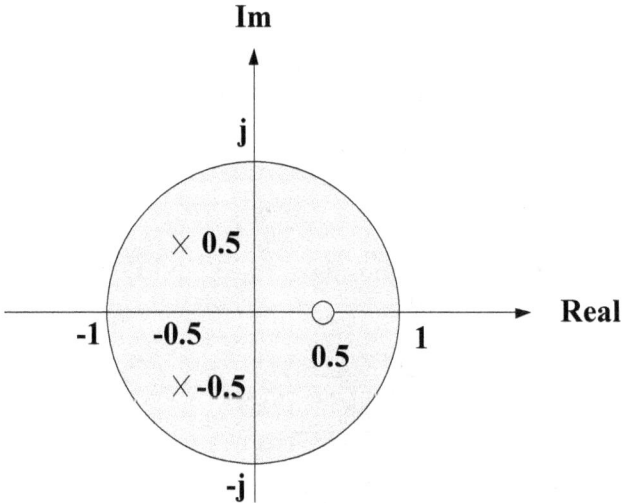

**Figure 10.9** Z-plane representation **of** Example 10.6 a

b- Putting the numerator polynomial equal to zero and then finding the zeros roots of the transfer function,

$$z^2 + 0.25 = 0$$
$$z^2 = -0.25$$
$$z = \sqrt{-0.25} = \pm j\, 0.5$$

Therefore, we get $z_1 = +j0.5$ and $z_2 = -j0.5$ as the roots.
Now, setting the denominator polynomial equal to zero and finding the roots gives us the poles of the transfer function,

$$(z - 0.75)(z^2 + 4z + 4.5) = 0$$

It leads to

$$z - 0.75 = 0 \rightarrow z = 0.75$$

and

$$z = \frac{-4 \pm \sqrt{(4)^2 - 4(1)(4.5)}}{2(1)} = -2 \pm j0.707$$

Therefore, the poles are $p_1 = 0.75$, $p_2 = -2 + j0.707$ and $p_3 = -2 - j0.707$. The magnitudes of these poles are
It can be noticed that the two poles outside the unit circle on the z-plane pole-zero
Therefore, the system is unstable. It is shown in the following figure.

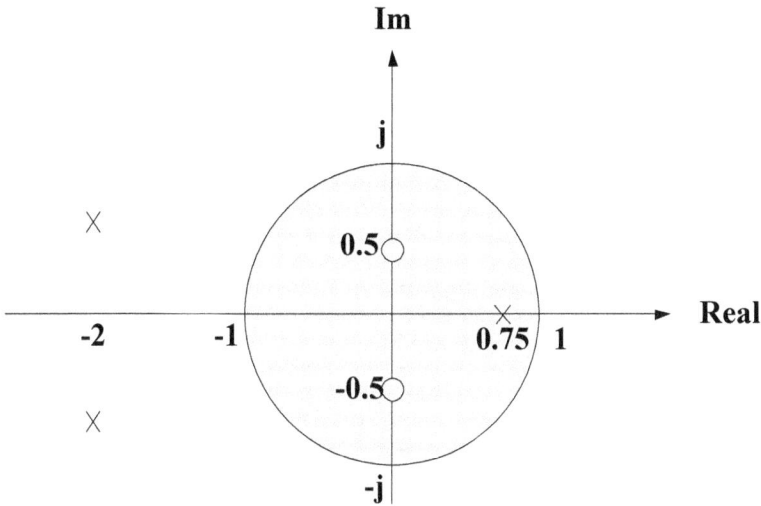

**Figure 10.10** Z-plane representation **of** Example 10.4 b

c- a- Putting the numerator polynomial equal to zero and then finding the roots,

$$z + 0.25 = 0$$

Therefore, we get $z_1 = -0.25$ as the root.

Now, setting the denominator polynomial equal to zero and finding the roots gives us the poles of the transfer function,

$$(z - 0.25)(z^2 + 1.5\,z + 1) = 0$$

This leads to

$$z - 0.5 = 0 \; so \; z = 0.25$$

and

$$z = \frac{-1.5 \pm \sqrt{(1.5)^2 - 4(1)(1)}}{2(1)} = -0.75 \pm j0.66$$

Therefore, the poles are $p_1 = 0.25$, $p_2 = -0.75 + j0.66$ and $p_3 = -0.75 - j0.66$. The magnitudes of these poles are

$$|p_1| = 0.25$$
$$|p_2| = |-0.75 + j0.66| = 1$$
$$|p_3| = |-0.75 - j0.66| = 1$$

It can be noticed that the magnitudes of the two poles are on unity circle 1, so the system is marginally stable. It is shown in the following figure.

d)

$(z^2 + z + 0.25)=0$ gives the two zeros at z=-0.5

$$(z - 1)^2(z + 1)(z - 0.6) = 0$$

It gives the poles at p=1 (two poles) and one pole at p=-1 and p=0.6, respectively

And the system is marginally stable

Example 10.7
    Check the stability of the system given by
$$H(z) = \frac{A(z - 1)^2}{(z - 0.4)(z - 0.5 + j0.5)(z - 0.5 - j0.5)}$$
Where A is a constant number.
Solution

$$\begin{aligned}
z - 0.4 = 0 &\rightarrow z = 0.4 \\
z - 0.5 + j0.5 = 0 &\rightarrow z = 0.5 - j0.5 \\
z - 0.5 - j0.5 = 0 &\rightarrow z = 0.5 + j0.5
\end{aligned}$$

The poles inside the unit circle, so the system is stable

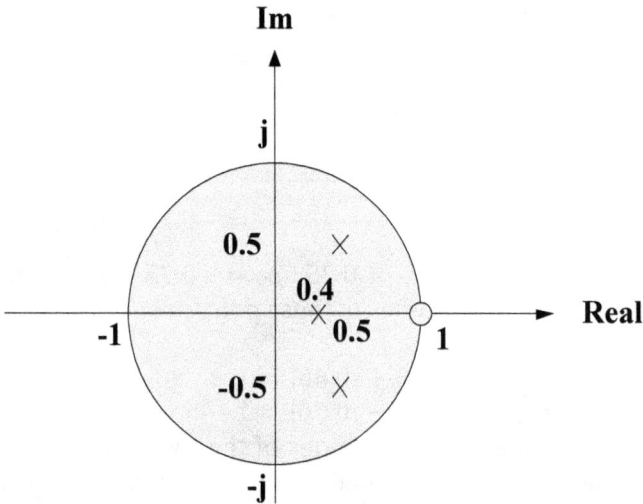

**Figure 10.11** Z-plane representation **of** Example 10.7

## Problems

10.1 Convert each of the following transfer functions into its pole-zero form

$$H(z) = \frac{z(1 - 0.16\,z^{-2})}{1 + 0.7\,z^{-1} + 0.1\,z^{-2}}$$

10.2 Given each of the following transfers, functions describe digital systems, sketch the z-plane pole-zero plot, and determine the stability status for the digital system.

(a)   $H(z) = \dfrac{z(z - 0.75)}{(z+0.25)(z^2 + z + 0.8)}$

(b)   $H(z) = \dfrac{z(z^2 + 0.25)}{(z-0.5)(z^2 + 4\,z + 7)}$

(c)   $H(z) = \dfrac{z(z + 0.15)}{(z+0.2)(z^2 + 1.4141\,z + 1)}$

(d)   $H(z) = \dfrac{z(z^2 + z + 0.25)}{(z-1)(z+1)^2(z-0.36)}$

10.3 An LTI system is represented by the difference equation:
y(n)=0.7 y(n-1) –0.12 y(n-2) +x(n-1) +x(n-2)
1-Find the system transfer function H(z).
2-Draw and obtain poles and zeros, is the system is stable?.
3-Find the output response if the input is the unit step sequence given by x(n)=u(n).

10.4 A DSP system is described by the following difference equation
$$y(n) = x(n) - 0.9x(n - 1) + 1.8x(n - 2))$$
Find the transfer function H(z), the Numerator polynomial A(z), and the denominator polynomial equation B(z).

10.5 A DSP system is described by the following difference equation
$$y(n) = x(n) - 1.2\,x(n - 1) + 0.7x(n - 2))$$
Find the transfer function H(z), the Numerator polynomial A(z), and the denominator polynomial equation B(z).

10.6
The following transfer functions describe digital systems

(a) $H(z) = \dfrac{(z - 1.5)}{(z+1.5)(z^2 + z + 0.5)}$

(b)   $H(z) = \dfrac{(z^2 + 1.25)}{(z-0.5)(z^2 + 4\,z +3)}$

(c)   $H(z) = \dfrac{(z + 0.25)}{(z-0.25)(z^2 +1.5\,z + 3)}$

(d)    $H(z) = \dfrac{(z^2 + z + 0.25)}{(z-1)^2(z+0.9)(z-0.2)}$

For each, sketch the z-plane pole-zero plot and determine the stability status for the digital system.

# References

1. python.org, "python.org - https://www.python.org," 2018.
2. python.org, "Python 3.7.1 documentation - https://docs.python.org/3.7/," 2018.
3. H.-P. Halvorsen, "Technology blog - https://www.halvorsen.blog," 2018.
4. Forman S. Acton. Numerical Methods that Work. Mathematical Association of America, Washington, D.C., 1990.
5. Joseph Adler. R in a Nutshell. O'Reilly, Sebastopol, CA, 2010.
6. Bernd Blasius, Amit Huppert, and Lewi Stone. Complex dynamics and phase synchronization in spatially extended ecological systems. Nature, 399:354–359, 1999.
7. Victor Bloomfield. Computer Simulation and Data Analysis in Molecular Biology and Biophysics: An Introduction Using R. Springer, New York, 2009.
8. John Verzani. Using R for Introductory Statistics. CRC Press, Boca Raton, 2004.
9. Petre Teodorescu Nicolae-Doru Stanescu, Nicolae Pandrea, Numerical Analysis with Applications in Mechanics and Engineering, by The Institute of Electrical and Electronics Engineers, Inc. Published by John Wiley & Sons, Inc. 2013.
10. Jorn Anemuller, Terrence J. Sejnowski, and Scott Makeig. Complex independent component analysis of frequency-domain electroencephalographic data. Neural Networks, 16(9):1311–1323, November 2003.
11. S. Barnet. Matrices. Methods and Applications. Oxford Applied Mathematics and Computing Science Series. Clarendon Press, 1990.
12. Christopher Bishop. Neural Networks for Pattern Recognition. Oxford University Press, 1995.
13. Robert J. Boik. Lecture notes: Statistics 550. Online, April 22 2002. Notes.
14. D. H. Brandwood. A complex gradient operator and its application in adaptive array theory. IEE Proceedings, 130(1):11–16, February 1983. PTS. F and H.
15. M. Brookes. Matrix Reference Manual, 2004. Website May 20, 2004.
16. Mads Dyrholm. Some matrix results, 2004. Website August 23, 2004.

17. Nielsen F. A., Formula, Neuro Research Unit and Technical university of Denmark, 2002.
18. Gelman A. B., J. S. Carlin, H. S. Stern, D. B. Rubin, Bayesian Data Analysis, Chapman and Hall / CRC, 1995.
19. Gene H. Golub and Charles F. van Loan. Matrix Computations. The Johns Hopkins University Press, Baltimore, 3rd edition, 1996.
20. Trefethen, L. N., "Is Gauss quadrature better than Clenshaw-Curtis?" SIAM Rev., vol. 50, no. l,pp. 67-87,2008.
21. Trefethen, L. N., Spectral Methods in MATLAB, SIAM, Philadelphia, 2000.
22. Boyd, John, "Exponentially convergent Fourier-Chebyshev quadrature schemes on bounded and infinite intervals," J. Sei. Comput., vol. 2, pp. 99-109, 1987.
23. Trefethen, L. N., and Bau, David, Numerical Linear Algebra, SIAM, Philadelphia, 1997.
24. Watkins, David, Fundamentals of Matrix Computations, John Wiley & Sons, Inc., New York, 3 rd Edition, 2010.
25. Braun, Martin, Differential Equations and Their Applications, Springer-Verlag, New York, 1993 (4th Edition).
26. Burden, Richard, and Faires, J. Douglas, Numerical Analysis, Brooks/Cole, Pacific Grove, CA, 1997 (6th Edition).
27. Brigham, E. O., *The Fast Fourier Transform and Its Applications*. Englewood Cliffs, NJ: Prentice Hall, 1988.
28. Samir I. Abood, Digital Signal Processing: A Primer With MATLAB® 1st Edition, CRC Press, Taylor and Francis Group, 2020.

# Appendix A

# Mathematical Formulas

The appendix contains all the formulas wanted to solve the problems in this book.

## A.1 Quadratic Formulas

The roots of the quadratic equation $ax^2 + bx + c = 0$

$$x_1, x_2 = \frac{-b \pm \sqrt{b^2 - 4ac}}{2a}$$

## A.2 Trigonometric Identities

$$\sin(-x) = -\sin x$$

$$\cos(-x) = \cos x$$

$$\sec x = \frac{1}{\cos x}, \qquad \csc x = \frac{1}{\sin x}$$

$$\tan x = \frac{\sin x}{\cos x}, \qquad \cot x = \frac{1}{\tan x}$$

$$\sin(x \pm 90°) = \pm \cos x$$

$$\cos(x \pm 90°) = \mp \sin x$$

$$\sin(x \pm 180°) = -\sin x$$

$$\cos(x \pm 180°) = -\cos x$$

$$\cos^2 x + \sin^2 x = 1$$

$$\frac{a}{\sin A} = \frac{b}{\sin B} = \frac{c}{\sin C} \qquad \text{(law of sines)}$$

$$a^2 = b^2 + c^2 - 2bc \cos A \qquad \text{(law of cosines)}$$

$$\frac{\tan\frac{1}{2}(A-B)}{\tan\frac{1}{2}(A+B)} = \frac{a-b}{a+b} \qquad \text{(law of tangents)}$$

$$\sin(x \pm y) = \sin x \cos y \pm \cos x \sin y$$

$$\cos(x \pm y) = \cos x \cos y \mp \sin x \sin y$$

$$\tan(x \pm y) = \frac{\tan x \pm \tan y}{1 \mp \tan x \tan y}$$

$$2\sin x \sin y = \cos(x-y) - \cos(x+y)$$

$$2\sin x \cos y = \sin(x+y) - \sin(x-y)$$

$$2\cos x \cos y = \cos(x+y) - \cos(x-y)$$

$$\sin 2x = 2\sin x \cos x$$

$$\cos 2x = \cos^2 x - \sin^2 x = 2\cos^2 x - 1 = 1 - 2\sin^2 x$$

$$\tan 2x = \frac{2\tan x}{1 - \tan^2 x}$$

$$\cos^2 x = \frac{1 + \cos(2x)}{2}$$

$$\sin^2 x = \frac{1 - \cos(2x)}{2}$$

$$a\cos x + b\sin x = K\cos(x+\theta), \quad \text{where } K = \sqrt{a^2 + b^2} \text{ and } \theta = \tan^{-1}\left(\frac{-b}{a}\right)$$

$$e^{\pm jx} = \cos x \pm j\sin x \qquad \text{(Euler's formula)}$$

$$\cos x = \frac{e^{jx} + e^{-jx}}{2}$$

$$\sin x = \frac{e^{jx} - e^{-jx}}{2j}$$

## A.3 Trigonometric Substitution

| Form | Trig Sub | Identity |
|---|---|---|
| $\sqrt{a^2 + x^2}$ | $x = a\tan\theta$ | $1 + \tan^2\theta = \sec^2\theta$ |
| $\sqrt{a^2 - x^2}$ | $x = a\sin\theta$ | $1 - \sin^2\theta = \cos^2\theta$ |
| $\sqrt{x^2 - a^2}$ | $x = a\sec\theta$ | $\sec^2\theta - 1 = \tan^2\theta$ |

## A.4 Hyberbolic Functions

$$\sinh x = \frac{1}{2}\left(e^x - e^{-x}\right)$$

$$\cosh x = \frac{1}{2}\left(e^x + e^{-x}\right)$$

$$\tanh x = \frac{\sinh x}{\cosh x}$$

$$\coth x = \frac{1}{\tanh x}$$

$$\operatorname{csch} x = \frac{1}{\sinh x}$$

$$\operatorname{sech} x = \frac{1}{\cosh x}$$

$$\sinh(x \pm y) = \sinh x \cosh y \pm \cosh x \sinh y$$
$$\cosh(x \pm y) = \cosh x \cosh y \pm \sinh x \sinh y$$
$$\tan(x \pm y) = \frac{\tan x \pm \tan y}{1 \mp \tan x \tan y}$$

## A.5 Derivatives

If $U = U(x)$, $V = V(x)$, and $a$ = constant,

$$\frac{d}{dx}(aU) = a\frac{dU}{dx}$$

$$\frac{d}{dx}(UV) = U\frac{dV}{dx} + V\frac{dU}{dx}$$

273

$$\frac{d}{dx}\left(\frac{U}{V}\right) = \frac{V\dfrac{dU}{dx} - U\dfrac{dV}{dx}}{V^2}$$

$$\frac{d}{dx}\left(aU^n\right) = naU^{n-1}$$

$$\frac{d}{dx}\left(a^U\right) = a^U \ln a \frac{dU}{dx}$$

$$\frac{d}{dx}\left(e^U\right) = e^U \frac{dU}{dx}$$

$$\frac{d}{dx}\left(\sin U\right) = \cos U \frac{dU}{dx}$$

$$\frac{d}{dx}\left(\cos U\right) = -\sin U \frac{dU}{dx}$$

$$\frac{d}{dx}\tan U = \frac{1}{\cos^2 U}\frac{dU}{dx}$$

## A.6 Indefinite Integrals

If $U = U(x)$, $V = V(x)$, and $a$ = constant,

$$\int a\,dx = ax + C$$

$$\int U\,dV = UV - \int V\,dU \quad \text{(integration by parts)}$$

$$\int U^n\,dU = \frac{U^{n+1}}{n+1} + C, \quad n \neq 1$$

$$\int \frac{dU}{U} = \ln U + C$$

$$\int a^U\,dU = \frac{a^U}{\ln a} + C, \quad a > 0, a \neq 1$$

$$\int e^{ax}\,dx = \frac{1}{a}e^{ax} + C$$

$$\int xe^{ax}\,dx = \frac{e^{ax}}{a^2}(ax - 1) + C$$

$$\int x^2 e^{ax}\,dx = \frac{e^{ax}}{a^3}\left(a^2 x^2 - 2ax + 2\right) + C$$

$$\int 1nxdx = x1nx - x + C$$

$$\int \sin axdx = -\frac{1}{a}\cos ax + C$$

$$\int \cos axdx = \frac{1}{a}\sin ax + C$$

$$\int \sin^2 axdx = \frac{x}{2} - \frac{\sin 2ax}{4a} + C$$

$$\int \cos^2 axdx = \frac{x}{2} + \frac{\sin 2ax}{4a} + C$$

$$\int x\sin axdx = \frac{1}{a^2}(\sin ax - ax\cos ax) + C$$

$$\int x\cos axdx = \frac{1}{a^2}(\cos ax + ax\sin ax) + C$$

$$\int x^2 \sin axdx = \frac{1}{a^3}(2ax\sin ax + 2\cos ax - a^2x^2\cos ax) + C$$

$$\int x^2 \cos axdx = \frac{1}{a^3}(2ax\cos ax - 2\sin ax + a^2x^2\sin ax) + C$$

$$\int e^{ax}\sin bxdx = \frac{e^{ax}}{a^2 + b^2}(a\sin bx - b\cos bx) + C$$

$$\int e^{ax}\cos bxdx = \frac{e^{ax}}{a^2 + b^2}(a\cos bx + b\sin bx) + C$$

$$\int \sin ax\sin bxdx = \frac{\sin(a-b)x}{2(a-b)} - \frac{\sin(a+b)x}{2(a+b)} + C, \quad a^2 \neq b^2$$

$$\int \sin ax\cos bxdx = -\frac{\cos(a-b)x}{2(a-b)} - \frac{\cos(a+b)x}{2(a+b)} + C, \quad a^2 \neq b^2$$

$$\int \cos ax \cos bx dx = \frac{\sin(a-b)x}{2(a-b)} + \frac{\sin(a+b)x}{2(a+b)} + C, \quad a^2 \neq b^2$$

$$\int \frac{dx}{a^2 + x^2} = \frac{1}{a} \tan^{-1} \frac{x}{a} + C$$

$$\int \frac{x^2 dx}{a^2 + x^2} = x - a \tan^{-1} \frac{x}{a} + C$$

$$\int \frac{dx}{\left(a^2 + x^2\right)^2} = \frac{1}{2a^2} \left( \frac{x}{x^2 + a^2} + \frac{1}{a} \tan^{-1} \frac{x}{a} \right) + C$$

## A.7 Definite Integrals

If $m$ and $n$ are integers,

$$\int_0^{2\pi} \sin ax \, dx = 0$$

$$\int_0^{2\pi} \cos ax \, dx = 0$$

$$\int_0^{\pi} \sin^2 ax \, dx = \int_0^{\pi} \cos^2 ax dx = \frac{\pi}{2}$$

$$\int_0^{\pi} \sin mx \sin nx dx = \int_0^{\pi} \cos mx \cos nx dx = 0, \quad m \neq n$$

$$\int_0^{\pi} \sin mx \cos nx dx = \begin{cases} 0, & m+n = \text{even} \\ \dfrac{2m}{m^2 - n^2}, & m+n = \text{odd} \end{cases}$$

$$\int_0^{2\pi} \sin mx \sin nx dx = \int_{-\pi}^{\pi} \sin mx \sin nx dx = \begin{cases} 0, & m \neq n \\ \pi, & m \neq n \end{cases}$$

276

$$\int_0^\infty \frac{\sin ax}{x} dx = \begin{cases} \dfrac{\pi}{2}, & a > 0 \\ 0, & a = 0 \\ -\dfrac{\pi}{2}, & a < 0 \end{cases}$$

$$\int_0^\infty \frac{\sin^2 x}{x} dx = \frac{\pi}{2}$$

$$\int_0^\infty \frac{\cos bx}{x^2 + a^2} dx = \frac{\pi}{2a} e^{-ab}, \quad a > 0, b > 0$$

$$\int_0^\infty \frac{x \sin bx}{x^2 + a^2} dx = \frac{\pi}{2} e^{-ab}, \quad a > 0, b > 0$$

$$\int_0^\infty \sin cx dx = \int_0^\infty \sin c^2 x dx = \frac{1}{2}$$

$$\int_0^\pi \sin^2 nx dx = \int_0^\pi \sin^2 x dx = \int_0^\pi \cos^2 nx dx = \int_0^\pi \cos^2 x dx = \frac{\pi}{2}, \quad n = \text{an integer}$$

$$\int_0^\pi \sin mx \sin nx dx = \int_0^\pi \cos mx \cos nx dx = 0, \quad m \neq n, m, n \text{ integers}$$

$$\int_0^\pi \sin mx \cos nx dx = \begin{cases} \dfrac{2m}{m^2 - n^2}, & m + n = \text{odd} \\ 0, & m + n = \text{even} \end{cases}$$

$$\int_{-\infty}^\infty e^{\pm j 2\pi t x} dx = \delta(t)$$

$$\int_0^\infty x^n e^{-ax} dx = \frac{n!}{a^{n+1}}$$

$$\int_0^\infty e^{-a^2 x^2} dx = \frac{\sqrt{\pi}}{2a}, \quad a > 0$$

$$\int_0^\infty x^{2n} e^{-ax^2}\, dx = \frac{1 \bullet 3 \bullet 5 \bullet \bullet \bullet (2n-1)}{2^{n+1} a^n} \sqrt{\frac{\pi}{a}}$$

$$\int_0^\infty x^{2n+1} e^{-ax^2}\, dx = \frac{n!}{2a^{n+1}}, \qquad a > 0$$

## A.8 L'Hopital's Rule

If $f(0) = 0 = h(0)$, then

$$\lim_{x\to 0} \frac{f(x)}{h(x)} = \lim_{x\to 0} \frac{f'(x)}{h'(x)}$$

where the prime indicates differentiation.

## A.9 Summation

$$\sum_{k=1}^{N} k = \frac{1}{2} N(N+1)$$

$$\sum_{k=1}^{N} k^2 = \frac{1}{6} N(N+1)(2N+1)$$

$$\sum_{k=1}^{N} k^3 = \frac{1}{4} N^2 (N+1)^2$$

$$\sum_{k=0}^{N} a^k = \frac{a^{N+1} - 1}{a-1} \qquad a \neq 1$$

$$\sum_{k=M}^{N} a^k = \frac{a^{N+1} - a^M}{a-1} \qquad a \neq 1$$

$$\sum_{k=0}^{N} \binom{N}{k} a^{N-k} b^k = (a+b)^N, \text{ where } \binom{N}{k} = \frac{N!}{(N-k)!k!}$$

## A.10 Numerical Integration Approximations

$$TRAP(n) = \frac{b-a}{2n} \left[ f(x_0) + 2f(x_1) + 2f(x_2) + \ldots + 2f(x_{n-1}) + f(x_n) \right]$$

$$SIMP(n) = \frac{b-a}{3n}\left[f(x_0) + 4f(x_1) + 2f(x_2) + 4f(x_3) + \ldots + 2f(x_{n-2}) + 4f(x_{n-1}) + f(x_n)\right]$$

## A.11 Powers of the Trig Functions

1) Integrals of the form: $\int \sin^m x \cos^n x\, dx$   m or n odd

Strategy:  If m is odd, save a sine factor and convert to cosine
If n is odd, save a cosine factor and convert to sine

2) Integrals of the form: $\int \sin^m x \cos^n x\, dx$   m and n even and non-negative

Strategy: use ½ angle identities:
$$\cos^2 x = \frac{1 + \cos(2x)}{2}; \quad \sin^2 x = \frac{1 - \cos(2x)}{2}$$

3) Integrals of the form: $\int \sec^m x \tan^n x\, dx$ ; if m is even

Strategy: Save a $\sec^2 x$ and convert to tangent

4) Integrals of the form: $\int \sec^m x \tan^n x\, dx$ ; if n is odd

Strategy:  Save a secx tanx and convert to secant

5) $\int \tan^n x\, dx$ n any positive integer

Strategy:  convert a tan² x to sec²x-1 and distribute; repeat if necessary

6) $\int \sec^m x\, dx$ ; m is odd

Strategy:  Integrate by parts

# Appendix B

## Table B.1 Laplace Transforms of some Basic Functions

| $f(t)$ | $L\,(f(t))$ | $f(t)$ | $L\,(f(t))$ |
|---|---|---|---|
| $1$ | $\dfrac{1}{s}$ | $\cos^2 kt$ | $\dfrac{s^2 + 2k^2}{s(s^2 + 4k^2)}$ |
| $t$ | $\dfrac{1}{s^2}$ | $e^{at}$ | $\dfrac{1}{s-a}$ |
| $t^n$ | $\dfrac{n!}{s^{n+1}}$ | $\sinh k\,t$ | $\dfrac{k}{s^2 - k^2}$ |
| $t^{-1/2}$ | $\sqrt{\dfrac{\pi}{s}}$ | $\cosh kt$ | $\dfrac{s}{s^2 - k^2}$ |
| $t^{1/2}$ | $\dfrac{\sqrt{\pi}}{2s^{3/2}}$ | $\sinh^2 kt$ | $\dfrac{2k^2}{s(s^2 - 4k^2)}$ |
| $\sin k\,t$ | $\dfrac{k}{s^2 + k^2}$ | $\cosh^2 kt$ | $\dfrac{s^2 - 2k^2}{s(s^2 - 4k^2)}$ |
| $\cos k\,t$ | $\dfrac{s}{s^2 + k^2}$ | $t.\,e^{at}$ | $\dfrac{1}{(s-a)^2}$ |
| $\sin^2 kt$ | $\dfrac{2k^2}{s(s^2 + 4k^2)}$ | $t^n.\,e^{at}$ $n$ a positive integer | $\dfrac{n!}{(s-a)^{n+1}}$ |
| $e^{at}\sin kt$ | $\dfrac{k}{(s-a)^2 + k^2}$ | $H\,(t-a) = u_a(t)$ | $\dfrac{e^{-as}}{s}, s > 0$ |
| $e^{at}\cos kt$ | $\dfrac{s-a}{(s-a)^2 + k^2}$ | $\delta(t)$ | $1$ |
| $e^{at}\sinh kt$ | $\dfrac{k}{(s-a)^2 - k^2}$ | $\delta(t-t_0)$ | $e^{-st_0}$ |
| $e^{at}\cosh kt$ | $\dfrac{s-a}{(s-a)^2 - k^2}$ | $e^{at} f(t)$ | $F(s-a)$ |
| $t \sin kt$ | $\dfrac{2ks}{y(s^2 + k^2)^2}$ | $f(t-a)\,H(t-a)$ | $e^{-as} F(s)$ |
| $t \cos kt$ | $\dfrac{s^2 - k^2}{(s^2 + k^2)^2}$ | $f^{(n)}(t)$ | $s^n F(s) - s^{n-1} f(0) \ldots\ldots - f^{(n-1)}(0)$ |
| $\sin kt + kt \cos kt$ | $\dfrac{2ks^2}{(s^2 + k^2)^2}$ | $t^n f(t)$ | $(-1)^n \dfrac{d^n}{ds^n} F(s)$ |

| | | | |
|---|---|---|---|
| $sin\ kt$ $-\ kt\ cos\ kt$ | $\dfrac{2k^3}{(s^2+k^2)^2}$ | $\int_0^t f\quad (u)$ $.g(t-u)du$ | $F(s)\ G(s)$ |
| $t\ Sinh\ k\ t$ | $\dfrac{2ks}{(s^2-k^2)^2}$ | $\dfrac{sin\ at}{t}$ | $arc\ tan\left(\dfrac{a}{s}\right)$ |
| $t\ cosh\ kt$ | $\dfrac{s^2+k^2}{(s^2-k^2)^2}$ | $\dfrac{1}{\sqrt{\pi t}}e^{-a^2/4t}$ | $\dfrac{e^{-a\sqrt{s}}}{\sqrt{s}}$ |
| $\dfrac{e^{at}-e^{bt}}{a-b}$ | $\dfrac{1}{(s-a)(s-b)}$ | $\dfrac{e^{at}-e^{bt}}{t}$ | $ln\dfrac{s\text{-}a}{s-b}$ |
| $\dfrac{ae^{at}-e^{bt}}{a-b}$ | $\dfrac{s}{(s-a)(s-b)}$ | $1-cos\ kt$ | $\dfrac{k^2}{s(s^2+k^2)}$ |

# Appendix C

## FOURIER SERIES

This Appendix is concerned with a means of analyzing systems with periodic excitations. Fourier series, a premier tool for analyzing periodic signals. Fourier series is named after Jean Baptiste Joseph Fourier (1768-1830), a French mathematician and physicist. In 1822, Fourier was the first to suggest that any periodic function can be represented as a sum of sinusoids. Such a representation provides a powerful tool for analyzing signals and systems. System responses to periodic signals are of practical interest because these signals are common.

Fourier analysis (Fourier series and Fourier transform) plays a major role in system analysis. Fourier analysis leads to the frequency spectrum of a continuous-time signal. The frequency spectrum displays the various sinusoidal components that make up the signal. Engineers think of signals in terms of their frequency spectra and of systems in terms of their frequency response. Fourier analysis converts time-domain signals into frequency-domain representation which lends new insight into the nature and the properties of the signals and systems. For many purposes, the frequency-domain representations are more convenient to analyze, synthesize, and process. The frequency domain linear systems are described by linear algebraic equations which can be easily solved, in contrast to the time domain representation, where they are described by linear differential equations. It is for these reasons that Fourier analysis is used extensively today in science and engineering.

### C.1 TRIGONOMETRIC FOURIER SERIES

The Fourier series can be represented in three ways, the sine-cosine, amplitude-phase, and complex exponential.

A periodic signal is one that repeats itself every T seconds. In other words, a continuous time signal $x(t)$ satisfies
$$x(t) = x(t + nT)$$

$$(C.1)$$

where n is an integer and T is the fundamental period of x(t).

Any periodic function can be expressed as an infinite series consisting of sine or cosine functions. Thus, x(t) can be expressed as

$$x(t) = a_0 + a_1 cos\omega_0 t + b_1 sin\omega_0 t + a_2 cos2\omega_0 t + b_2 sin2\omega_0 t + a_3 cos3\omega_0 t + b_3 sin3\omega_0 t + \cdots \qquad (C.2)$$

This can be written as

$$x(t) = a_0 + \sum_{n=1}^{\infty}(a_n \cos n\omega_0 t + b_n \sin n\omega_0 t) \qquad (C.3)$$

$$\omega_0 = \frac{2\pi}{T} \qquad (C.4)$$

$\omega_0$ is known as the fundamental frequency (in rad/sec). The coefficients $a$'s are called the Fourier cosine coefficients including $a_0$, the constant term, which is in reality the 0-th cosine term, and b's are called
the Fourier sine coefficients. The Fourier coefficient $a_0$ in the above equation is the constant or dc component of x(t). The Fourier coefficients $a_n$ and $b_n$ (for $n \neq 0$) are the amplitudes of the sinusoids in the ac component of x(t).

A periodic function x(t) can be expanded as a Fourier series only if it fulfills the Dirichlet conditions given as:
1. x(t) should be integrable over any period; that is,
    . $\int_{t_0}^{t_0+T} |x(t)| dt < \infty$
2. x(t) has only a finite number of maxima and minima over any period.
3. x(t) has only a finite number of discontinuities over any period.

The process of determining the Fourier coefficients a0, an, and bn is called Fourier analysis. The coefficients can be determined as follows.

$$a_0 = \frac{1}{T} \int_0^T x(t)dt \qquad (C.5a)$$

$$a_n = \frac{2}{T} \int_0^T x(t)\cos n\omega_0 t dt \qquad (C.5b)$$

284

$$b_n = \frac{2}{T} \int_0^T x(t) \sin n\omega_0 t \, dt \qquad\qquad (C.5c)$$

The fact that sine and cosine functions are orthogonal over a period T leads to the below trigonometric integrals.

$$\int_0^T \sin n\omega_0 t \, dt = 0 = \int_0^T \cos n\omega_0 t \, dt \qquad\qquad (C.6a)$$

$$\int_0^T \sin n\omega_0 t \cos n\omega_0 t \, dt = 0 \qquad\qquad (C.6b)$$

$$\int_0^T \sin n\omega_0 t \sin m\omega_0 t \, dt = 0 =$$
$$\int_0^T \cos n\omega_0 t \cos m\omega_0 t \, dt, \qquad m \neq n \qquad\qquad (C.6c)$$

$$\int_0^T \sin^2 n\omega_0 t \, dt = \frac{T}{2} = \int_0^T \cos^2 n\omega_0 t \, dt \qquad\qquad (C.6d)$$

The above integral will be used in finding the Fourier coefficients.

To find $a_0$, integrate both sides of Equation (C.3) over one period.

$$\int_0^T x(t) \, dt = \int_0^T a_0 \, dt + \sum_{n=1}^{\infty} \left[ \int_0^T a_n \cos n\omega_0 t \, dt \right] +$$
$$\sum_{n=1}^{\infty} \left[ \int_0^T b_n \sin n\omega_0 t \, dt \right] \qquad\qquad (C.7)$$

$$= a_0 T + 0 + 0$$

Using the identities in Equation (C.6a), the two integrals having the ac terms becomes zero. Therefore

$$\int_0^T x(t) \, dt = a_0 T$$

$$a_0 = \frac{1}{T} \int_0^T x(t) \, dt \qquad\qquad (C.8)$$

To find $a_n$, multiply both sides of Equation (C.3) by $\cos m\omega_0 t$ and integrate over one period.

$$\int_0^T x(t) \cos m\omega_0 t\, dt$$

$$= \int_0^T a_0 \cos m\omega_0 t\, dt$$

$$+ \sum_{n=1}^{\infty} \left[ \int_0^T a_n \cos n\omega_0 t \cos m\omega_0 t\, dt \right]$$

$$+ \sum_{n=1}^{\infty} \left[ \int_0^T b_n \sin n\omega_0 t \cos m\omega_0 t\, dt \right] \qquad (C.9)$$

$$= 0 + \frac{T}{2} a_n + 0$$

The integral containing, $a_0$ and $b_n$ becomes zero according to eqs. (C.6a), and (C.6b) respectively. The integral containing $a_n$ will be zero except when m=n in which case, it is T/2 according to Equation (C.6d). Thus,

$$\int_0^T x(t) \cos m\omega_0 t\, dt = a_n \frac{T}{2}, \quad m = n$$

or

$$a_n = \frac{2}{T} \int_0^T x(t) \cos n\omega_0 t\, dt \qquad (C.10)$$

To find $b_n$, multiply both sides of Equation (C.3) by $\sin m\omega_0 t$ and integrate over one period.

$$\int_0^T x(t) \sin m\omega_0 t dt$$

$$= \int_0^T a_0 \sin m\omega_0 t dt$$

$$+ \sum_{n=1}^{\infty} \left[ \int_0^T a_n \cos n\omega_0 t \sin m\omega_0 t dt \right]$$

$$+ \sum_{n=1}^{\infty} \left[ \int_0^T b_n \sin n\omega_0 t \sin m\omega_0 t dt \right] \qquad (C.11)$$

$$= 0 + 0 + \frac{T}{2} b_n$$

The integral containing $a_0$ and $a_n$ becomes zero according to eqs. (C.6a), and (C.6b) respectively. The integral containing $b_n$ will be zero except when m=n in which case, it is T/2 according to Equation (C.6d). Thus,

$$\int_0^T x(t) \sin n\omega_0 t dt = b_n \frac{T}{2}, \qquad m = n$$

or
$$b_n = \frac{2}{T} \int_0^T x(t) \sin n\omega_0 t dt \qquad (C.12)$$

Since x(t) is periodic the integration of Equation (C.5) over one full period from $t_0$ to $t_0 + T$ or from $-T/2$ to $T/2$ instead of 0 to T, gives the same results.

Equation (C.3) is the sine-cosine form of Fourier series. An alternative representation is the amplitude-phase (or polar) form:

$$x(t) = a_0 + \sum_{n=1}^{\infty} A_n \cos(n\omega_0 t + \phi_n) \qquad (C.13)$$

This combines the sine-cosine pair at frequency $n\omega_o$ into a single sinusoid. We can apply the trigonometric identity

$$\cos(A + B) = cosA\ cosB - sinA\ sinB \quad (C.14)$$

to the ac terms in Equation (C.13) to get

$$a_0 + \sum_{n=1}^{\infty} A_n \cos(n\omega_0 t + \emptyset_n)$$

$$= a_0$$
$$+ \sum_{n=1}^{\infty} (A_n \cos \emptyset_n) \cos n\omega_0 t$$
$$+ \sum_{n=1}^{\infty} (-A_n \sin \emptyset_n) \sin n\omega_0 t$$

When we equate the coefficients of the series expansions in Equation (C.3) and (C.15), we obtain

$$a_n = A_n cos\emptyset_n, \quad b_n = -A_n \sin \emptyset_n \quad (C.16a)$$

$$A_n = \sqrt{a_n^2 + b_n^2}, \emptyset_n = -\tan^{-1}\frac{b_n}{a_n} \quad (C.\,16b)$$

These may also be related in a compact, complex form as

$$A_n \angle \emptyset_n = a_n - jb_n \quad (C.17)$$

The frequency components of the signal x(t) can be displayed in terms of the amplitude and phase spectra. The frequency spectrum of x(t) is the combination of both the amplitude and phase spectra.

## C.2 EXPONENTIAL FOURIER SERIES

We can now see that the sine and cosine functions can be represented in terms of complex exponentials. It turns out that we can use complex exponentials to represent Fourier series. In many respects, this makes for a simpler representation.

The Fourier series representation of x(t), can be expressed in complex exponential form as

$$x(t) = \sum_{n=-\infty}^{\infty} c_n e^{jn\omega_0 t} \tag{C.18}$$

where $\omega_0 = 2\pi/T$ is the fundamental frequency and the coefficients $c_n$ are given by

$$c_n = \frac{1}{T} \int_0^T x(t) e^{-jn\omega_0 t} dt \tag{C.19}$$

By substituting the following Euler's identities in Equation (C.3)

$$\cos n\omega_0 t = \frac{1}{2} \left[ e^{jn\omega_0 t} + e^{-jn\omega_0 t} \right] \tag{C.20a}$$

$$\sin n\omega_0 t = \frac{1}{2j} \left[ e^{jn\omega_0 t} - e^{-jn\omega_0 t} \right] = -\frac{j}{2} \left[ e^{jn\omega_0 t} - e^{-jn\omega_0 t} \right] \tag{C.20b}$$

We get,

$$x(t) = a_0 + \frac{1}{2} \sum_{n=1}^{\infty} a_n \left( e^{jn\omega_0 t} + e^{-jn\omega_0 t} \right)$$
$$- j\, b_n \left( e^{jn\omega_0 t} - e^{-jn\omega_0 t} \right)$$

$$= a_0 + \frac{1}{2} \sum_{n=1}^{\infty} \left[ (a_n - jb_n) e^{jn\omega_0 t} \right.$$
$$\left. + (a_n + jb_n) e^{-jn\omega_0 t} \right] \tag{C.21}$$

So the new coefficient $c_n$ will be

$$c_0 = a_0,$$

$$c_n = \frac{(a_n - jb_n)}{2} = |c_n| \angle \theta_n, \tag{C.22}$$

289

$$c_{-n} = c_n^* = \frac{(a_n + jb_n)}{2}$$

so that x(t) becomes

$$x(t) = c_0 + \sum_{n=1}^{\infty} \left[ c_n e^{jn\omega_0 t} + c_{-n} e^{-jn\omega_0 t} \right] \qquad (C.23a)$$

The concise form is

$$x(t) = \sum_{n=-\infty}^{\infty} c_n e^{jn\omega_0 t}$$

$$= \sum_{n=-\infty}^{\infty} |c_n| e^{j(n\omega_0 t + \theta_n)} \qquad (C.23b)$$

The complex Fourier coefficients $c_n$ can be readily obtained as follows using eqs.(C.5b) and (C.5c) for $a_n$, $b_n$.

$$c_0 = a_0 = \frac{1}{T} \int_0^T x(t) dt$$

$$(C.24a)$$

For n = 1, 2, 3, .... we have

$$c_n = \frac{(a_n - jb_n)}{2} = \frac{1}{T} \int_0^T x(t) (\cos n\omega_0 t - j\sin n\omega_0 t) dt =$$
$$\frac{1}{T} \int_0^T x(t) e^{-jn\omega_0 t} dt \qquad (C.24b)$$

Similarly

$$c_{-n} = c_n^* = \frac{(a_n + jb_n)}{2} = \frac{1}{T} \int_0^T x(t) e^{jn\omega_0 t} dt$$

The last expression is equivalent to stating that for n = -1, -2, -3,......

$$c_n = \frac{1}{T} \int_0^T x(t) e^{-jn\omega_0 t} dt \qquad (C.24c)$$

The three eqs. (C.24a), (C.24b), (C.24c) can be combined into one expression

$$c_n = \frac{1}{T}\int_0^T x(t)e^{-jn\omega_0 t}dt \qquad \text{for} \qquad n = 0,\pm1,\pm2,\pm3,\ldots\ldots \qquad \text{(C.25)}$$

Therefore the complex Fourier series is

$$x(t) = \sum_{n=-\infty}^{\infty} c_n e^{jn\omega_0 t} \qquad \text{(C.26)}$$

Equation (C.26) is also called complex frequency spectrum of x(t) which is formed by both the complex amplitude spectrum (even symmetry) and the phase spectrum (odd symmetry).

## C.3 PROPERTIES OF FOURIER SERIES

Some properties of Fourier series are discussed in this section. These properties provide us a better understanding of the Fourier series.

### Linearity

The Fourier series expansion for periodic signals x(t) and y(t) with the same period is given by

$$x(t) = \sum_{n=-\infty}^{\infty} \alpha_n \exp[jn\omega_0 t] \qquad \text{(C. 27)}$$

$$y(t) = \sum_{n=-\infty}^{\infty} \beta_n \exp[jn\omega_0 t] \qquad \text{(C. 28)}$$

If z(t) is a linear combination of x(t) and y(t), then

$$z(t) = k_1 x(t) + k_2 y(t) \qquad \text{(C. 29)}$$

where $k_1$ and $k_2$ are arbitrary constants. Then we can write,

$$z(t) = \sum_{n=-\infty}^{\infty} (k_1 \alpha_n + k_2 \beta_n) \exp[jn\omega_0 t]$$

$$= \sum_{n=-\infty}^{\infty} \gamma_n \exp[jn\omega_0 t] \qquad \text{(C. 30)}$$

This implies that the Fourier coefficients are

$$\gamma_n = k_1 \alpha_n + k_2 \beta_n \qquad \text{(C.31)}$$

**Time shifting**

When a time shift is applied to a periodic signal x(t), the period T of the signal is preserved. The Fourier series coefficient $\gamma_n$ of the resulting signal x(t-τ) may be expressed as

$$\gamma_n = \frac{1}{T} \int_0^T x(t - \tau) \exp[-jn\omega_0 t]\, dt =$$
$$\exp[-j\,\omega_0\tau] \frac{1}{T} \int_0^T x(\lambda) \exp[-jn\omega_0\lambda]\, d\lambda \qquad \text{(C.32)}$$

$$= \alpha_n \exp[-jn\omega_0\tau]$$

Where $\alpha_n$ is the $n^{th}$ Fourier series coefficient of x(t). Here we have changed variables by letting λ= t - τ.  That is, if

$$x(t) \xrightarrow{\phantom{xx}FS\phantom{xx}} \alpha_n$$

then

$$x(t - \tau) \xrightarrow{\phantom{xx}FS\phantom{xx}} \alpha_n e^{-jn\omega_0\tau}$$

$$\text{(C.33)}$$

One consequence of this property is that, when a periodic signal is shifted in time, the magnitude of the Fourier coefficient remains unaltered, that is $|\gamma_n| = |\alpha_n|$.

## Time reversal

Time reversal property, when applied to a continuous-time signal results in a time reversal of the corresponding sequence of Fourier series coefficients. To determine the Fourier series coefficients of $y(t) = x(-t)$, consider

$$x(-t) = \sum_{n=-\infty}^{\infty} \alpha_n \exp\left[-\frac{jn\pi t}{T}\right]$$

(C.34)

By substituting n = - m, we get

$$y(t) = x(-t) = \sum_{m=-\infty}^{\infty} \alpha_{-m} \exp\left[\frac{jm\pi t}{T}\right]$$

(C.35)

Thus, we conclude that

$$x(t) \xrightarrow{FS} \alpha_n$$

$$x(-t) \xrightarrow{FS} \alpha_{-n}$$

(C.36)

An interesting consequence of time reversal is that when x(t) is even, that is, if x(-t) = x(t) then its Fourier series coefficients are also even, that is $\alpha_{-n} = \alpha_n$. Similarly when x(t) is odd, that is, if x(-t) = - x(t) then its Fourier series coefficients are also odd, that is $\alpha_{-n} = -\alpha_n$.

## Time scaling

Time scaling operation applies directly to each of the harmonic component of x(t). If x(t) has the Fourier series representation $x(t) = \sum_{n=-\infty}^{\infty} \alpha_n \exp[jn\omega_0 t]$, then the Fourier series representation of the time-scaled signal x(at) is

$$x(at) = \sum_{n=-\infty}^{\infty} \alpha_n \exp[jn\omega_0 at] \qquad (C.37)$$

i.e.

$$x(at) \xrightarrow{\text{FS}} \alpha_n \tag{C.38}$$

The Fourier series coefficients have not changed, the Fourier series representation has changed because of the change in the fundamental frequency $a\omega_0$.

## Even and Odd Symmetries

Three types of symmetries have been discussed - even, odd and half-wave odd symmetry.

A function $x(t)$ is even if $x(t) = x(-t)$, and odd if $x(t) = - x(-t)$. Any function $x(t)$ is a sum of an even function and an odd function, and this can be done in only one way.

We know that any periodic function $x(t)$, with period $2\pi$, has a Fourier expansion of the form

$$x(t) = a_0 + \sum_{n=1}^{\infty} a_n \cos(n\omega_0 t) + b_n \sin(n\omega_0 t) \tag{C.39}$$

If $x(t)$ is even then all the $b_n$ 's vanish and the Fourier series is simply

$$x(t) = a_0 + \sum_{n=1}^{\infty} a_n \cos(n\omega_0 t) \tag{C.40}$$

If $x(t)$ is odd then all the $a_0$, $a_n$ 's vanish and the Fourier series is simply

$$x(t) = \sum_{n=1}^{\infty} b_n \sin(n\omega_0 t) \tag{C.41}$$

A periodic function possesses half wave symmetry if it satisfies the constraints

$$x(t) = -x(t - \frac{T}{2}) \tag{C.42}$$

If a function is shifted one half period and inverted, it looks identical to the original then it is called a half wave symmetry. A half wave odd symmetry can be even, odd or neither.

$$
\boxed{
\begin{aligned}
a_0 &= 0, \qquad a_n = 0 \; and \quad b_n = 0 \qquad for \; n \; even \\[2mm]
a_n &= \frac{4}{T} \int_0^{\frac{T}{2}} x(t) \cos n\omega_0 t \, dt, \qquad\qquad for \; n \; odd \\[2mm]
b_n &= \frac{4}{T} \int_0^{\frac{T}{2}} x(t) \sin n\omega_0 t \, dt, \qquad\qquad for \; n \; odd
\end{aligned}
}
$$

(C.43)

This shows that the Fourier series of a half-wave symmetric function contains only odd harmonics.

**Parseval's Theorem**

Parseval's theorem states that if x(t) is a periodic function with period T , then the average power P of the signal is defined by

$$P = \frac{1}{T} \int_0^T x^2(t) \, dt \tag{C.44}$$

Again let x(t) be an arbitrary periodic signal with period T, and consider the Fourier series of x(t) given by Equation (C.18). By Parseval's theorem, the average power P of the signal x(t) is given by

$$P = \sum_{n=-\infty}^{\infty} |c_n|^2 \tag{C.45}$$

In other words, the theorem states that the mean square value of the signal x(t) over one period equals the sum of the squared magnitudes of all the complex Fourier coefficients.

**Proof of the Parseval's theorem**

Assume x(t) has a complex Fourier series of the usual form:

$$x(t) = \sum_{n=-\infty}^{\infty} c_n e^{jn\omega_0 t} \; , \qquad \omega_0 = \left(\frac{2\pi}{T}\right)$$

$$c_n = \frac{1}{T} \int_0^T x(t) e^{-jn\omega_0 t}\, dt$$

$$x^2(t) = x(t)x(t) = x(t) \sum c_n e^{jn\omega_0 t}$$
$$= \sum c_n\, x(t)\, e^{jn\omega_0 t}\, dt$$

$$P = \frac{1}{T} \int_0^T x^2(t)\, dt = \frac{1}{T} \int_0^T \sum c_n\, x(t)\, e^{jn\omega_0 t}\, dt$$

$$= \frac{1}{T} \sum c_n \int_0^T x(t) e^{jn\omega_0 t}\, dt$$

$$= \sum c_n \cdot c_n^*$$

$$P = \sum_{n=-\infty}^{\infty} |c_n|^2 \qquad\qquad (C.46)$$

Parseval's theorem can also be written in terms of the Fourier coefficients $a_n$, $b_n$ of the trigonometric Fourier series. The engineering interpretation of this theorem is as follows (suppose x(t) denotes an electrical signal (current or voltage)) then from elementary circuit theory $x^2(t)$ is the instantaneous power (in a 1 ohm resistor) so that

$$P = \frac{1}{T} \int_0^T x^2(t)\, dt \qquad\qquad (C.47)$$

is the energy dissipated in the resistor during one period.

Table: Power associated with Fourier series coefficients

| Term | Sine-cosine form | Amplitude-phase form | Exponential form |
|------|------------------|----------------------|------------------|
| dc term | $a_0^2$ | $a_0^2$ | $|c_0|^2$ |
| others $|c_{-n}|^2$ | $\frac{1}{2}(a_n^2 + b_n^2)$ | $\frac{1}{2} A_n$ | $2|c_n|^2 = |c_n|^2 +$ |

## Truncated Complex Fourier Series

The Fourier series representation of x(t) can be expressed in complex exponential form as

$$x(t) = \sum_{n=-\infty}^{\infty} c_n e^{jn\omega_0 t} \qquad (C.48)$$

It is infinite and to truncate x(t) to finite partial sum of $x_N(t)$ at n = N, we can write Equation(C.48) as follows

$$x_N(t) = \sum_{n=-N}^{N} c_n e^{jn\omega_0 t} \qquad (C.49)$$

## C.4 FOURIER TRANSFORM

The Fourier transform is similar to Laplace transform. Both are integral transforms. The Fourier transform may be regarded a special case of Laplace transform with s=jω, when the Laplace transform exists. Just like the bilateral Laplace transform, Fourier transform can deal with systems having inputs for t < 0 as well as those for t >0.

if x(t) is a periodic function with period T, then we can expand it in a complex Fourier Series,

$$x(t) = \sum_{n=-\infty}^{\infty} c_n e^{jn\omega_0 t} \qquad (1)$$

Where

$$c_n = \frac{1}{T} \int_{-\frac{T}{2}}^{\frac{T}{2}} x(t) e^{-jn\omega_0 t} dt \qquad (2)$$

The spacing between adjacent harmonics is

$$\Delta\omega = (n+1)\omega_0 - n\omega_0 = \omega_0 = \frac{2\pi}{T} \qquad (3)$$

We substitute eq. (2) into eq. (1) and putting $\frac{1}{T} = \frac{\omega_0}{2\pi}$ we get

$$x(t) = \sum_{n=-\infty}^{\infty} \left[ \frac{1}{T} \int_{-\frac{T}{2}}^{\frac{T}{2}} x(t) e^{-jn\omega_0 t} dt \right] e^{jn\omega_0 t}$$

297

$$= \sum_{n=-\infty}^{\infty} \left[ \frac{\Delta\omega}{2\pi} \int_{-\frac{T}{2}}^{\frac{T}{2}} x(t) \, e^{-jn\omega_0 t} \, dt \right] e^{jn\omega_0 t} \quad (4)$$

$$= \frac{1}{2\pi} \sum_{n=-\infty}^{\infty} \left[ \int_{-\frac{T}{2}}^{\frac{T}{2}} x(t) \, e^{-jn\omega_0 t} \, dt \right] \Delta\omega e^{jn\omega_0 t}$$

In view of the discussion above, as T → ∞, we can put $\omega_0$ as $\Delta\omega$ and replace the sum over the discrete frequencies $n\omega_0$ by an integral over all frequencies. We replace $n\omega_0$ by a general frequency variable $\omega$. We then obtain the double integral representation

$$x(t) = \frac{1}{2\pi} \int_{-\infty}^{\infty} \left[ \int_{-\infty}^{\infty} x(t) e^{-j\omega t} \, dt \right] e^{j\omega t} \, d\omega \quad (5)$$

The inner integral is known as the Fourier transform of x(t) and is represented by X(ω),

$$X(\omega) = \mathscr{F}[x(t)] = \int_{-\infty}^{\infty} x(t) \, e^{-j\omega t} \, dt \quad (6)$$

the inverse Fourier transform as

$$x(t) = \mathscr{F}^{-1}[X(\omega)] = \frac{1}{2\pi} \int_{-\infty}^{\infty} X(\omega) \, e^{j\omega t} \, d\omega$$

**Table: Properties of the Fourier transform**

| Property | x(t) | X(ω) |
|---|---|---|
| 1. Linearity | $a_1 x_1(t) + a_2 x_2(t)$ | $a_1 X_1(\omega) + a_2 X_2(\omega)$ |
| 2. Scaling | x(at) | $\frac{1}{|a|} X\left(\frac{\omega}{a}\right)$ |
| 3. Time shift | x(t-a) | $e^{-j\omega a} X(\omega)$ |
| 4. Frequency shift | $e^{j\omega_0 t} x(t)$ | $X(\omega - \omega_0)$ |
| 5. Modulation | $\cos(\omega_0 t) x(t)$ | $\frac{1}{2}[X(\omega + \omega_0) + X(\omega - \omega_0)]$ |
| | $\sin(\omega_0 t) x(t)$ | $\frac{j}{2}[X(\omega + \omega_0) - X(\omega - \omega_0)]$ |
| 6. Time differentiation | $\frac{dx}{dt}$ | $j\omega X(\omega)$ |

298

$$\frac{d^n x}{dt^n} \qquad\qquad (j\omega)^n X(\omega)$$

7. Frequency differentiation $\quad (-jt)^n x(t) \qquad \dfrac{d^n}{d\omega^n} X(\omega)$

8. Time integration $\quad \int_{-\infty}^{t} x(t)dt \qquad \dfrac{X(\omega)}{j\omega} + \pi X(0)\delta(\omega)$

9. Time reversal $\qquad\qquad x(-t) \qquad\qquad X(-\omega) \text{ or } X^*(\omega)$

10. Duality $\qquad\qquad\quad X(t) \qquad\qquad 2\pi x(-\omega)$

11. Time convolution $\qquad x_1(t) * x_2(t) \qquad X_1(\omega)X_2(\omega)$

12. Frequency convolution $\; x_1(t)x_2(t) \qquad \dfrac{1}{2\pi}X_1(\omega) * X_2(\omega)$

13. Parseval's relation $\;\int_{-\infty}^{\infty} |x(t)|^2 dt \quad \dfrac{1}{2\pi}\int_{-\infty}^{\infty} |X(\omega)|^2 d\omega$

---

## Table: Fourier transform pairs.

| $x(t)$ | $X(\omega)$ |
|---|---|
| 1. $\delta(t)$ | $1$ |
| 2. $1$ | $2\pi\delta(\omega)$ |
| 3. $u(t)$ | $\pi\delta(\omega) + \dfrac{1}{j\omega}$ |
| 4. $u(t+\tau) - u(t-\tau)$ | $2\dfrac{\sin \omega\tau}{\omega}$ |
| 5. $|t|$ | $\dfrac{-2}{\omega^2}$ |
| 6. $\text{sgn}(t)$ | $\dfrac{2}{j\omega}$ |
| 7. $e^{-at}u(t)$ | $\dfrac{1}{a+j\omega}$ |
| 8. $e^{at}u(-t)$ | $\dfrac{1}{a-j\omega}$ |
| 9. $t^n e^{-at}u(t)$ | $\dfrac{n!}{(a+j\omega)^{n+1}}$ |
| 10. $e^{-a|t|}$ | $\dfrac{2a}{a^2+\omega^2}$ |
| 11. $|t|e^{-at} u(t)$ | $\dfrac{4aj\omega}{a^2+\omega^2}$ |
| 12. $e^{j\omega_0 t}$ | $2\pi\delta(\omega - \omega_0)$ |
| 13. $\sin \omega_0 t$ | $j\pi[\delta(\omega + \omega_0) - \delta(\omega - \omega_0)]$ |
| 14. $\cos \omega_0 t$ | $\pi[\delta(\omega + \omega_0) - \delta(\omega - \omega_0)]$ |
| 15. $(\sin \omega_0 t)u(t)$ | $j\pi[\delta(\omega + \omega_0) - \delta(\omega - \omega_0)] + \dfrac{\omega_0}{\omega_0^2-\omega^2}$ |
| 16. $(\cos \omega_0 t)u(t)$ | $\pi[\delta(\omega + \omega_0) - \delta(\omega - \omega_0)] + \dfrac{j\omega}{\omega_0^2-\omega^2}$ |
| 17. $\sin(\omega_0 t+\theta)$ | $j\pi[e^{-j\theta}\delta(\omega + \omega_0) - e^{j\theta}\delta(\omega - \omega_0)]$ |

18. $\cos(\omega_0 t+\theta)$  $\pi[e^{-j\theta}\delta(\omega+\omega_0)+e^{j\theta}\delta(\omega-\omega_0)]$

19. $e^{-at}\sin \omega_o\, tu(t)$  $\dfrac{\omega_0}{(a+j\omega)^2+\omega_0^2}$

20. $e^{-at}\cos \omega_o\, tu(t)$  $\dfrac{a+j\omega}{(a+j\omega)^2+\omega_0^2}$

21. $\Pi(t/\tau)$  $\tau \sin c\left(\dfrac{\omega\tau}{2}\right)$

22. $\Lambda\left(\dfrac{t}{\tau}\right)$  $\tau \sin c^2\left(\dfrac{\omega\tau}{2}\right)$

23. $\sum_{n=-\infty}^{\infty}\delta(t-nT)$  $\omega_0\sum_{n=-\infty}^{\infty}\delta(\omega-n\omega_0)$